RAND

How Americans View World Population Issues

A Survey of Public Opinion

David M. Adamson
Nancy Belden
Julie DaVanzo
Sally Patterson

Supported by the
David and Lucile Packard Foundation
William and Flora Hewlett Foundation
Rockefeller Foundation

POPULATION MATTERS

A RAND Program of Policy-Relevant Research Communication

PREFACE

This report presents the results of a survey of Americans' views about global demographic trends and issues in the context of U.S. international economic assistance. The survey was conducted in August–September 1998 by Belden Russonello & Stewart, a public-opinion research firm in Washington, D.C., in conjunction with Ronald Hinckley of Research/Strategy/Management, and coordinated by Sally Patterson of Wagner Associates Public Affairs Consulting, Inc., a public-affairs consulting firm based in Washington, D.C., and New York, N.Y. The work was performed jointly for RAND, with funding from the David and Lucile Packard Foundation, and the Centre for Development and Population Activities (CEDPA).

The survey explored public knowledge about world population facts and trends and public views on specific issues, such as U.S. international economic assistance, family planning, and abortion. This report should be of interest to anyone concerned with what the public knows and thinks about population-related issues.

RAND's portion of this research was conducted for the *Population Matters* project within RAND's Labor and Population Program. A principal goal of *Population Matters* is to inform public and elite audiences about findings of demographic research and their implications for policy. We also hope to identify information gaps and misperceptions and determine which of these can effectively be filled by scientific information. A related report explores congressional views on population issues (*How Does Congress Approach Population and Family Planning Issues? Results of Qualitative Interviews with Legislative Directors*, RAND, MR-1048-WFHF/RF/UNFPA).

The *Population Matters* project is funded by grants from the William and Flora Hewlett Foundation, the Rockefeller Foundation, and the David and Lucile Packard Foundation. For further information on the *Population Matters* project, contact:

Julie DaVanzo, Director, *Population Matters*
RAND
P.O. Box 2138
1700 Main St.
Santa Monica, CA 90407-2138
Julie_DaVanzo@rand.org

Or visit the project's website at http://www.rand.org/popmatters

CONTENTS

FIGURES

TABLES

This report presents results of a survey of how the American public views global demographic trends and issues in the context of U.S. international economic assistance.

BACKGROUND

Prior to this survey, no comparable survey had been undertaken since 1994. During the interval (between 1994 and August 1998, when this survey was conducted), important changes have taken place. First, the International Conference on Population and Development (ICPD) was held in Cairo, Egypt, in late 1994. This conference provided a forum for articulating and codifying the relatively recent shift in international population policy away from concerns about overall population growth and aggregate statistics toward a more individual-level concern with reproductive health and freedom in achieving desired family size. Second, the Republican Party won control of Congress, bringing with it a focus on domestic issues and national security. The new majority brought into key decision-making positions a political faction that traditionally has not placed a high priority on international development assistance. Third, the end of the Cold War in 1991 triggered a continuing and thorough reassessment of America's role in the world. Calls for a continued or even increased global role for the United States were countered by new isolationist sentiments.

Given these changes, a great deal of interest has arisen in policy and research communities about whether and to what extent American views of international population issues have changed.

To address these questions, the *Population Matters* project commissioned a nationally representative survey. The survey had two main purposes: to gain a clearer understanding of what the American public knows about global demographic trends and U.S. international economic assistance and to assess public attitudes and levels of concern about specific issues, such as family planning and abortion.

The survey was conducted by Belden Russonello & Stewart (BR&S), a public-opinion research firm in Washington, D.C. Telephone interviews were conducted with 1,500 U.S. residents aged 16 or older. When weighted to adjust for differences between our sample and the U.S. population in age, gender, and race, the survey data yield a nationally representative sample of Americans. The 1998 survey repeats a number of questions asked in a 1994 survey on these issues to allow for tracking attitude changes during the interim.

MAJOR FINDINGS

The survey asked about three groups of topics: attitudes about U.S. economic assistance overseas and priorities for targeting U.S. aid; knowledge and views of global demographic facts and trends; and views on specific issues, including family planning programs, abortion, and congressional actions on population-policy measures.

The survey found that, notwithstanding the perception that Americans may have become more absorbed with domestic problems since the end of the Cold War, a majority feels that U.S. economic assistance to other countries is necessary and appropriate. Most Americans lack knowledge about the world's population size and growth rate but are nonetheless concerned about global population growth and its possible consequences. Support for U.S. funding of voluntary family planning activities, both overseas and domestically, is overwhelming. The legal right to abortion, however, remains a contentious issue that divides the American public.

U.S. Economic Assistance Overseas

American support for international economic assistance is at its highest level since 1974, when opinions on the subject were first tracked. Almost six in ten respondents (59 percent) favor U.S. economic assistance to other countries. Moreover, support for international economic assistance is 50 percent or greater for every socioeconomic, demographic, and political subgroup we considered.

The top priorities for international economic assistance are health-related and humanitarian programs, including those aimed at improving health, child survival, the environment, and human rights, as well as reducing suffering and helping women in developing countries reduce the number of unintended pregnancies. Support is also high for programs that advance international relations goals, including promoting democracy, preventing war and conflict, promoting trade, and supporting friendly governments. Helping countries slow population growth ranks lower.

Public Knowledge and Attitudes About Global Demographic Trends

The American public is only mildly interested in demographic issues and has a limited sense of the current dimensions of world population. Only 14 percent know that world population size is about six billion. The same number think it is at least five times the current size as know the correct answer, and nearly 40 percent say they did not know the size of the world's population. Furthermore, most show little grasp of population growth rates. For example, nearly half say the world population will double in less than 20 years, when current demographic trends suggest that the doubling time will be around 50 years.

Notwithstanding this limited knowledge of population numbers, a majority of respondents believes that the world is overpopulated. Although the American public does not view rapid population growth as being as serious a world problem as disease and hunger, the spread of nuclear weapons, on threats to the environment, a majority believes that rapid population growth contributes to environmental problems, civil strife, and economic stagnation in developing countries.

A majority of those surveyed did not feel that low fertility in developed countries was a pressing population issue.

Specific Issues

Freedom to Achieve Desired Family Size. An overwhelming majority of the American public (92 percent) believes in the fundamental right of individuals and families to determine the number of children they will have and that the necessary means and information for accomplishing this should be available to all. This principle is the basic tenet underlying the ICPD program.

Family Planning Programs. Most Americans see voluntary family planning programs as necessary and beneficial. There is very strong support (86 percent) for the government providing family planning services to poor American women who want them, and an overwhelming majority of Americans (87 percent) favors requiring health insurers to cover the cost of contraception. More than three-fourths of every demographic and political subgroup we considered favor such programs. Eight in ten favor U.S. funding for voluntary family planning programs in developing countries. At least 70 percent of all demographic and political subgroups favor such funding. Most Americans believe that family planning services are not currently available to most people in the world.

Support for family planning in developing countries is related to the belief that it can reduce the number of abortions, that the world is overpopulated, and that rapid population growth is a serious problem, as well as to strong support for humanitarian and other economic assistance to developing countries. What limited opposition exists to U.S. funding for family planning programs in developing countries does not result from opposition to family planning in general, because at least two-thirds of those who oppose funding family planning abroad favor government and health insurers' support for family planning in the United States. The small opposition that does exist seems to stem from an opposition to economic assistance to other countries, a belief that rapid population growth is not a serious problem, and a belief that the availability of contraception encourages sexual activity among teenagers and unmarried couples.

Abortion. Attitudes about abortion in the United States have been remarkably stable over the last 25 years. Abortion remains an ever-present and divisive issue in the population policy arena. About half of the survey respondents opposed abortion either completely or except in cases of rape, incest, and danger to the mother's life. The other half supported a legal right to abortion. However, the public is not divisible simply into pro- and antiabortion rights segments because the great majority of Americans support or disapprove of abortion depending on circumstances.

Support for abortion appears to stem in part from a belief that legal abortion can save women's lives. Opposition to abortion appears to stem in part from belief that too many women use abortion as a routine means for controlling births and that the availability of legal abortion encourages sexual activity among teenagers and unmarried couples.

Attitudes toward abortion overseas are very similar to those regarding abortion in the United States—those who oppose one are likely to oppose the other. Fifty percent of respondents favor U.S. aid for voluntary, safe abortion as part of reproductive health care in developing countries that request it.

Relationship Between Family Planning and Abortion. As just discussed, 80 percent of our respondents favored U.S. funding for family planning programs in developing countries and 50 percent favored U.S. funding for voluntary safe abortion in developing countries that request it. We considered the overlap of these two groups. Of our entire sample, 45 percent favored funding for both family planning and abortion, 32 percent favored funding for family planning but opposed funding for abortion, while 14 percent opposed funding for both.   More than two-thirds of those who oppose funding for abortion in developing countries support funding for family planning in those countries. Those who favor support for family planning but oppose support for abortion are similar to those who favor funding for both in their support for international engagement and in their belief that improved availability of family planning can reduce abortion, but they are more likely to align with the group that opposes funding for both on all other issues regarding abortion. This suggests that an understanding of the potential of family planning to reduce abortion is associated with support for family planning.

The public is confused about whether the term "family planning" includes abortion. Forty-six percent of respondents said that "family planning" includes abortion, while 52 percent said that it does not.

For our other questions about support for voluntary international family planning programs, we stipulated that for purposes of the survey, "family planning" should be understood to exclude abortion.

Congressional Actions

The public has a mixed view of congressional actions regarding funding relating to population issues. Fifty percent of the public approves of the 1996 congressional vote to reduce the U.S. contribution to international family planning, while 45 percent disapprove of it. The apparent contradiction between this response and the strongly held belief that the United States should support family planning programs overseas suggests either that respondents are unaware of historic or current funding levels or that Americans support such programs in principle but are less supportive when it comes to funding them.

There is more disapproval of Congress's actions in preventing the United States from funding family planning in organizations that perform abortions, even if the U.S. contribution goes just for the family planning component of services. Here, 51 percent disapprove of Congress's denial of funding and 44 percent approve. When asked about U.N. dues, 48 percent disapproved of Congress's withholding a portion of the U.S. contribution to U.N. dues, while 36 percent approved.

IMPLICATIONS FOR RESEARCH COMMUNICATION

We draw six main implications for communicating demographic research findings to policy audiences and the general public.

First, few people are aware of the size and rate of growth of the world's population. Our survey suggests that focusing on aggregate numbers—as, for instance, much of the press coverage of the "Day of 6 Billion" in October of 1999 tended to do—is less likely to interest the public than a focus on individual perspectives. A focus on individual- and family-level quality-of-life issues, such as achieving desired family size, is consistent with the ICPD "approach" to framing population issues, although we cannot assess whether the ICPD has had any causal effect on American attitudes.

Second, the public does not see much of a connection between population trends and other issues, such as reproductive rights, the environment, or global security, and they care less about the former than the latter. Therefore, research communication could usefully emphasize the connections of population growth with high fertility and other issues. Research has shown, for example, strong links between women's fertility behavior and their own and their children's health. Indeed, additional research that explores the intersections of these areas, cutting across traditional fields of analysis, would be valuable in advancing public understanding of how demographic concerns relate to other issues thought to be more pressing.

Third, despite strong support for U.S. government funding for international family planning, half of the respondents did not oppose congressional cuts in funding for family planning programs in developing countries. Other research has shown that Americans tend to overestimate the fraction of the U.S. budget spent on foreign aid. They might also do this for family planning programs. In fact, funding for family planning programs is about 4.5 percent of foreign economic assistance. The public would benefit from accurate information about the relatively low cost of population assistance programs and the need for—as well as foreign governments' and individuals' continuing desire for—U.S. support for such programs.

Fourth, research shows that legal abortion can save women's lives, but only two-thirds of the overall population and only one-half of those who oppose U.S. support for abortion overseas recognize this.

Fifth, it is important that the public be informed about the potential of family planning to reduce abortion. Evidence from a number of countries (e.g., Russia, Kazakhstan, Bangladesh, Hungary, and South Korea) shows that improved availability of contraception has cut the number of abortions.

Finally, the public lacks a clear grasp of what the term "family planning" means and whether it encompasses abortion. This is not surprising, because the demographic research community itself does not seem to agree on a single definition of family planning. This finding implies that communicators should not always assume that their audiences know the meanings of terms like "family planning" or "birth control" and should define them whenever possible.

ACKNOWLEDGMENTS

The authors would like to acknowledge the contributions to the development of the survey and analyses and helpful commentary provided by Ronald Hinckley of Research/Strategy/Management; Julie Conrad at Belden Russonello & Stewart; Patricia Sears and Peggy Curlin at CEDPA; Ellen Marshall of the United Nations Foundation; Sarah Clark and the staff of the Population Program at the David and Lucile Packard Foundation; Joseph Speidel of the William and Flora Hewlett Foundation; Elizabeth Frankenberg, Lynn Karoly, and Michael Rich of RAND; and Priscilla Cavalca and Peter Farrand of Wagner Associates Public Affairs Consulting, Inc. Thanks are due also to Penny Mastt for her diligent secretarial support and to Dan Sheehan for his careful editing.

ACRONYMS

BR&S	Belden Russonello & Stewart
CEDPA	Centre for Development and Population Activities
FP	Family planning
ICPD	International Conference on Population and Development
RDD	Random digital dial

Chapter One

INTRODUCTION

This report presents the results of a comprehensive survey of American attitudes about population trends and issues, particularly in the context of U.S. economic assistance overseas. The survey is the first comprehensive attempt to gauge U.S. public attitudes on the subject since 1994.

PURPOSE AND MOTIVATION FOR THE SURVEY

This survey had two purposes: to gain a clearer understanding of what the American public knows about global demographic trends and related issues and to assess public attitudes about specific issues, such as international economic assistance, family planning, and abortion.

For several reasons, it was important to undertake this survey. First, current information is needed on how the public views U.S. global engagement with respect to population and development assistance. The most recent survey of public attitudes on these issues was conducted in 1994, just prior to the International Conference on Population and Development (ICPD), which was held in Cairo in September 1994. The ICPD was a multinational gathering sponsored by the United Nations that produced a comprehensive policy agenda and action items on a range of population issues. Signed by 178 nations, including the United States, the ICPD Programme of Action has influenced the way in which many stakeholder communities in the international arena frame population issues. Instead of focusing on aggregate population statistics and trends, these communities have become more focused on individual- and family-level quality-of-life issues, such as achieving desired family size and improving the health, education, and socioeconomic status of women and children. While we had no way of tracing any causal effect of the ICPD on public attitudes, we were interested in whether American attitudes had changed in a way that was consistent with the ICPD "approach" to framing population issues. To help us track these changes, this survey was designed to parallel the 1994 survey.

Prior to this survey, we conducted in-depth interviews with congressional staff on population-related topics (Patterson and Adamson, 1999). We discovered that the offices of congressional moderates— the approximately 10 percent of members of Congress who constitute the "swing vote" on virtually all population-related policy questions—were more interested in population issues and more sympathetic toward population-related program funding than was widely believed. We also discovered a wide range of opinion among even this small congressional subgroup about Americans' level of interest in the U.S. role abroad and U.S. leadership on international population issues such as family planning. The survey can help toward a better understanding of Americans' interest in these issues.

Finally, we also wanted to identify information gaps and misperceptions in Americans' understanding of these issues, especially those that could be addressed by the findings of scientific research. This information is important to the broader goals of the *Population Matters* project, which are to synthesize and communicate the findings and implications of demographic research in ways that policy analysts and others will find accessible. Because effectively informing these audiences requires understanding their attitudes and preconceptions, an important component of the project has been an assessment of audience knowledge and attitudes about population issues.

ORGANIZATION OF THE REPORT

In the next chapter, we discuss the survey methodology. Readers most interested in the findings may wish to proceed directly to Chapter Three, which presents findings on Americans' attitudes toward international economic assistance and priorities for allocating assistance. Within this context, we sought more specific information on how Americans view demographic issues. Chapter

Four presents findings on what Americans know about global demographic trends and the extent to which they view these trends as problems. Chapter Five discusses findings on specific demographic issues, including individuals' rights to achieve desired family size, family planning programs, and abortion. Chapter Six highlights key findings and conclusions and draws implications for efforts to communicate population research to the general public.

Chapter Two

METHODOLOGY

The survey was conducted by Belden Russonello & Stewart (BR&S), a public-opinion research firm in Washington, D.C., and coordinated by Sally Patterson of Wagner Associates Public Affairs Consulting, Inc., a New York and Washington, D.C.–based public-affairs consulting firm.[1] The 1998 survey repeats a number of questions asked in a 1994 survey on these issues (also conducted by Belden and Russonello [1994]) to allow for tracking attitude changes during the interim. A copy of the questionnaire for the 1998 survey, along with response totals for it and for the 1994 survey where comparable, is presented in Appendix A.

Telephone interviews were conducted in August-September of 1998 with 1,500 U.S. residents aged 16 or older.[2] When weighted to adjust for differences between our sample and the U.S. population in age, gender, and race, the survey data yield a nationally representative sample of Americans. Appendix B discusses methodological issues regarding the survey and analyses in more detail.

------

[1]Dr. Ronald Hinckley of Research/Strategy/Management contributed to the questionnaire construction and performed the factor and regression analyses.

[2]To achieve 1,500 completed interviews, 5,743 phone numbers were dialed. However, a number of these were never at home and/or may not have been eligible (once we were looking exclusively for 16–20-year-olds). Our ratio of completions-to-contacts is typical for a survey of this type. Our weighting procedure adjusts for age, sex, and racial/ethnic differences between our sample and the national population. Furthermore, survey research has shown that standard public opinion survey procedures typically produce substantive results virtually identical to the results of studies using more rigorous, and much more expensive, methods (Keeter and Miller, 1998).

SAMPLE

Most public opinion surveys are based on adults ages 18 and older. However, the sample for our 1998 survey on population and related issues was expanded to include 16- and 17-year-olds. Because of sponsor interest in the attitudes of young people and because young people represent a new challenge for population communicators, youths aged 16 to 20 were oversampled in our survey. Understanding their worldview and perceptions of population issues will help researchers present their findings more effectively to this age group.

The data collected in the survey have been weighted to adjust for differences between our sample and the U.S. population in age, gender, and race/ethnicity, to bring these demographic variables back to the proper proportion for the population and into their proper proportions within subgroups of age cohorts, racial/ethnic groups, and males and females. (See Table B.1 in Appendix B.) The statistics presented in this report are based on the weighted data.

Table 2.1 shows the socioeconomic, demographic, and political characteristics of the survey respondents. It presents characteristics shown subsequently in the report—those that generally yielded statistically significant variations in the answers to the questions related to overseas economic assistance and to population issues that we examine in Chapters Three, Four, and Five. The exact wording of the survey questions on these topics can be seen in the questionnaire in Appendix A.

PRESENTATION OF THE DATA

In the next three chapters we present tables and figures that show the responses to the survey. In the interest of brevity, some of the survey questions are summarized in the titles of the figures. The survey questions are available in their entirety in the appendixes. We do not present the questions in the same order that they were asked in the survey; the actual order used in the survey can be seen in Appendix A.

Table 2.1

Socioeconomic, Demographic, and Political
Characteristics of the Sample

	Unweighted Number	Unweighted Percentage	Weighted Percentage
Total	1,500	100	100
Gender			
Male	688	46	48
Female	812	54	52
Age			
16–20	200	13	9
21–29	211	14	16
30–44	457	30	31
45–59	324	22	23
60+	294	20	21
Race/Ethnicity			
White	1,181	79	75
African-American	144	10	11
Hispanic	100	7	10
Other	75	5	4
Education			
High school grad or less	672	45	43
Some college	362	24	25
College+	459	31	32
Household Income			
<$25,000	282	19	20
$25,000–$49,999	511	34	35
$50,000–$74,999	293	20	19
>$75,000	227	15	16
Party Affiliation[a]			
Democrat	524	35	36
Republican	428	29	29
Independent	393	26	26
Political Viewpoint			
Liberal	419	30	28
Moderate	422	28	28
Conservative	586	39	39
Religious Affiliation			
High Protestant[b]	315	21	21
Baptist	312	21	21
Other Protestant[c]	314	21	20
Catholic	362	24	24
Jewish	20	1	1
Muslim	3	0	1
No religious affiliation	128	9	9

Table 2.1—continued

	Unweighted Number	Unweighted Percentage	Weighted Percentage
Church Attendance			
More than once a week	189	13	12
About once a week	465	31	32
Attend less	836	56	56

NOTE: In some cases the unweighted numbers shown add to less than 1,500 and percentages add to less than 100 percent because of "don't know/refusal" answers.

[a]Registered voters.

[b]High Protestant = Episcopalian, Methodist, Congregational, Presbyterian, and Lutheran.

[c]Other Protestant = Pentecostal, Mormon, United Church of Christ, and others.

In some tables and figures, we show differentials by socioeconomic, demographic, political, and religious characteristics. The set of characteristics presented differs somewhat across the tables and figures. Each shows differentials by gender, age, and race/ethnicity and by other characteristics shown in Table 2.1 for which differentials were statistically significant.

In some figures, we compare responses in our 1998 survey to those to the same questions that were asked in a similar survey in 1994. Because the 1994 survey only included voters (i.e., those who voted in 1992) in its sample, in comparing results to the 1994 survey we show data for the 910 voters in the 1998 survey (i.e., those who voted in 1996) so that we are comparing similar samples.

FACTOR AND REGRESSION ANALYSES

The survey contains about 60 specific variables related to the various attitudes and opinions about international economic assistance and population issues ("topical variables"). Additionally, more than 30 demographic or behavioral variables identifying characteristics and activities are associated with each respondent. We have analyzed these data by examining each of the specific topical variables by each of the demographic or behavioral variables (a typical crosstab analysis). This provides great in-depth understanding of individual topical variables. However, it provides no understanding of any association that might exist *between* the topical variables or of the relative strength of any associations between the topical variables and the predictor (demographic and behavioral) variables.

We have used two steps to uncover the levels and degrees of association between variables. The first is *factor analysis,* which is a statistical procedure designed to identify and group together a large number of variables that are correlated or interrelated into a smaller number of sets. Each set of variables is called a "factor" and represents a unifying construct or concept derived from the nature of the individual variables that are interrelated. This procedure reduces the number of items under analysis and simplifies the description and understanding of otherwise complex and numerous phenomena. For example, instead of examining each individual tree in the forest (the 60 attitudinal variables in this data set), the trees are grouped according to shared characteristics (type of leaf or needle, bark, hardness or softness of wood, soil conditions for growth, etc.) into types of trees—oak, maple, pine, cedar, etc. (the 11 factors that emerged from the 60 variables). The forest of the survey data is then also understandable in terms of the underlying dimensions or factors in addition to an item-by-item basis.

The factor analysis procedure assigns a score for each respondent for each factor created. This creates a series of scales from high to low for each factor. We then conducted *multiple regression analyses* to examine correlates of each of these factor scales. The 30 or so predictor variables (socioeconomic, demographic, religious, and political characteristics of respondents) have been regressed against the attitudinal factor scores to determine the degree to which each predictor variable associates with each factor.

Additional details about the factor analyses and regression analysis methodologies can be found in Appendix B. The findings of the factor analyses are summarized in the ensuing chapters. A full discussion of the findings of both of these analyses appears in Appendix C.

Chapter Three

PUBLIC ATTITUDES TOWARD U.S. INTERNATIONAL ECONOMIC ASSISTANCE

Our first set of survey questions sought to update information on public attitudes about U.S. international involvement, specifically with respect to international economic assistance and priorities for allocating it. Some policy analysts have suggested that America is moving into a neo-isolationist period, and previous public opinion research has shown that Americans' concern for problems in the United States overshadows their sense of urgency about problems overseas.[1] This chapter looks at our findings regarding public support for various types of U.S. global economic assistance.

AMERICAN SUPPORT FOR U.S. INTERNATIONAL ECONOMIC ASSISTANCE

The current survey, taken during a period of economic prosperity in our own country, shows high levels of support for international economic assistance.

Q15. Are you generally in favor or opposed to the U.S. giving economic assistance to other countries? Is that very much (in favor/opposed) or somewhat (in favor/opposed)?

---

[1]For example, see Belden & Russonello, 1994; Doherty, 1993, p. 2267; Kull, 1995; Muravchik, 1996; and Yankelovich, 1991.

Six in ten (59 percent) say they favor the United States giving economic assistance to other countries (see Figure 3.1). Although the percentage favoring economic assistance outweighs those opposing by 59 percent to 37 percent, most do not have strongly held views on the subject. Two-thirds of those favoring economic assistance say they support it only somewhat.

As Table 3.1 illustrates, support for international economic assistance differs among various groups. Support is strongest among college graduates (73 percent favoring), African-Americans (69 percent), Democrats (67 percent), those with household incomes over $75,000 (65 percent), and liberals (64 percent). Nonetheless, support is 50 percent or greater for *every* subgroup shown in Table 3.1.

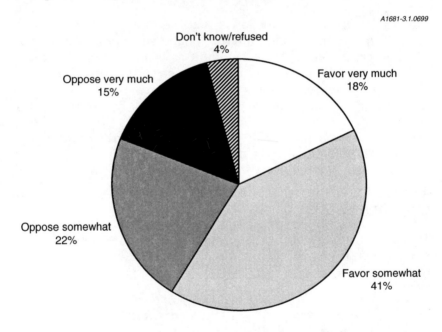

A1681-3.1.0699

Figure 3.1—Do You Favor or Oppose U.S. Economic Assistance
to Other Countries?

Table 3.1

**Opinions Regarding Economic Assistance to Other Countries,
by Socioeconomic and Demographic Characteristics**

	% Favor	% Oppose	% Favor Strongly	% Favor Some-what	% Oppose Some-what	% Oppose Strongly
Total	59	37	18	41	22	15
Gender						
Male	56	40	16	40	22	18
Female	61	35	19	42	22	13
Age						
16–20	64	31	22	42	22	9
21–29	60	37	17	44	25	12
30–44	60	35	16	44	20	15
45–59	55	42	20	35	28	15
60+	58	37	17	41	16	22
Race/Ethnicity						
White	57	39	15	42	23	16
African-American	69	27	26	43	15	12
Hispanic	61	36	26	34	23	13
Education						
High school grad or less	50	46	13	37	26	20
Some college	56	39	16	40	23	16
College+	73	25	25	48	16	8
Household Income						
<$25,000	52	43	12	40	27	16
$25,000–$49,999	61	37	17	43	21	16
$50,000–$74,999	57	39	18	39	21	18
>$75,000	65	32	28	36	19	13
Party Affiliation[a]						
Democrat	67	31	24	43	16	14
Republican	51	46	12	40	28	17
Independent	57	38	17	40	25	13
Political Viewpoint						
Liberal	64	34	25	38	21	13
Moderate	61	35	19	42	22	13
Conservative	54	42	12	42	24	19

[a]Registered voters.

It is important to keep in mind when considering the answers that show approval of international economic assistance and (as we will see later) high levels of support for the concept of family planning assistance that such support does not necessarily mean approval to spend at higher levels, or even at current levels.

TRENDS IN SUPPORT FOR ECONOMIC ASSISTANCE

Public-opinion surveys have been asking Americans about their support or opposition to U.S. economic assistance overseas since the early 1970s. The 59 percent level of support expressed in our 1998 survey is the highest since opinions on the subject were first tracked. The graph in Figure 3.2 tracks responses to the question of favoring or opposing economic assistance from a variety of surveys over the last 25 years. The low point measured in polling was in 1993, when less than half (43 percent) of the public said they favored U.S. economic assistance abroad.

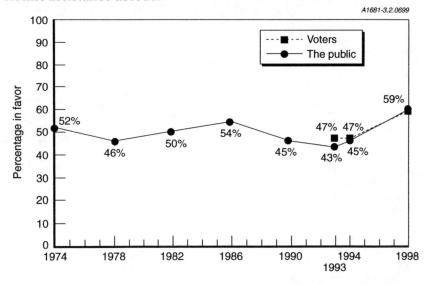

NOTE: The 1974, 1978, 1982, 1990, and 1994 data come from the Chicago Council on Foreign Relations (1995). The 1986 data are from Belden (1986). The 1993 data are from Belden & Russonello (1993). The 1998 data are derived from the survey that is the basis for this report. All the studies used the terminology "economic aid" or "economic assistance to other countries or nations."

Figure 3.2—Do You Favor or Oppose Giving International Economic Assistance to Other Countries?

VALUE OF INTERNATIONAL ECONOMIC ASSISTANCE TO THE UNITED STATES

Q42. Do you agree or disagree: Money the United States spends help-ing people overseas eventually helps the U.S. economically? Is that strongly or somewhat?

Sixty-one percent of the respondents agreed with this statement (see Figure 3.3). The factor analysis shows that approval of economic assistance (Q15) is associated with the belief that what we spend helping people overseas eventually helps the United States economi-cally. When people believe we in the United States are getting some-thing out of international economic aid, they are more likely to sup-port that funding.

A1681-3.3.0699

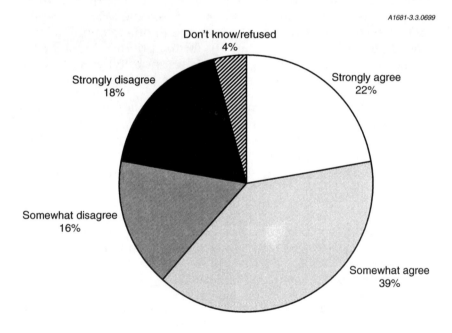

Figure 3.3—Does Money the United States Spends Overseas
Help the United States Economically?

PRIORITIZING GOALS FOR INTERNATIONAL ECONOMIC ASSISTANCE

We next inquired in the survey about a variety of types of possible goals for international economic assistance. The public approves of U.S. assistance being spent in a variety of ways. However, some goals are more strongly supported than others are.

Q16–30. Thinking about where you would like to see the U.S. use its resources, here are some possible goals for U.S. government assistance programs to other countries. Using a scale where 1 means lowest priority and 10 means top priority, please rate these goals for U.S. government assistance overseas.

Specifically, we asked the public to think about "where you would like to see the United States use its resources," and to score 15 possible goals for U.S. government assistance to other countries, on a scale of "1" for lowest priority to "10" for top priority. In Figure 3.4 we plot the percentage saying "10" (highest priority) and show, on the right-hand side, the average values on the 1–10 scale.

Two of the top goals address the well-being of children—the goals *of improving children's health* and *increasing child survival rates.* Improving children's health receives an average score of 7.8 out of a possible 10, with nearly four in ten giving it a "10" for top priority. Child survival has an average score of 7.4, with 33 percent giving this goal a score of 10.

Protecting the environment and relieving human suffering were also high priorities. *Environmental protection,* at an average of 7.7, is one of the most popular aims of U.S. international economic assistance; it was scored a "10" by 36 percent of the public. *Relieving suffering* was scored a "10" by 30 percent, and has an average score of 7.4.

Other goals that follow a little lower on the list include avoiding unintended pregnancy and efforts aimed at women's well-being. *Avoiding unintended pregnancy* receives an average of 6.9 and a "10" from 30 percent. Addressing the needs of women also receives relatively strong levels of interest but at a lower rate than the child-focused goals. *Improving women's health* receives an average score of 6.9 (with 25 percent ranking this goal a "10") and *improving the status of women* obtains a 6.5 (with 19 percent saying "10").

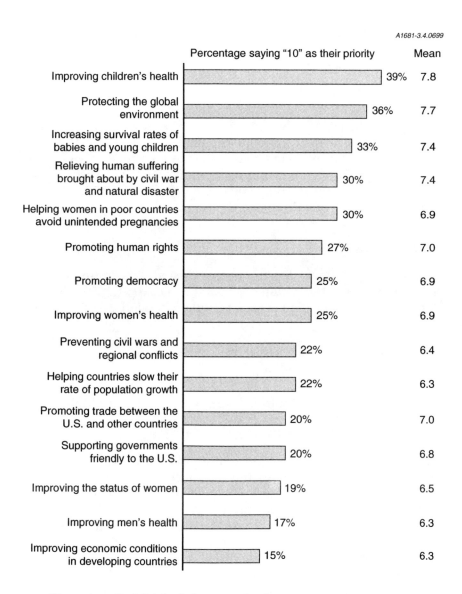

A1681-3.4.0699

	Percentage saying "10" as their priority	Mean
Improving children's health	39%	7.8
Protecting the global environment	36%	7.7
Increasing survival rates of babies and young children	33%	7.4
Relieving human suffering brought about by civil war and natural disaster	30%	7.4
Helping women in poor countries avoid unintended pregnancies	30%	6.9
Promoting human rights	27%	7.0
Promoting democracy	25%	6.9
Improving women's health	25%	6.9
Preventing civil wars and regional conflicts	22%	6.4
Helping countries slow their rate of population growth	22%	6.3
Promoting trade between the U.S. and other countries	20%	7.0
Supporting governments friendly to the U.S.	20%	6.8
Improving the status of women	19%	6.5
Improving men's health	17%	6.3
Improving economic conditions in developing countries	15%	6.3

Figure 3.4—To Which of These Goals of U.S. Economic Assistance Would You Assign High Priority?

Finally, the goal of addressing population growth receives a good score, but is lower priority than a number of others. Twenty-two percent rate *slowing rates of population growth* in other countries at

"10," and it obtains an average score of 6.3 (which, along with improving men's health and improving economic conditions in developing countries, is the lowest average score in the figure).

In sum, the data reveal that child-oriented goals for economic assistance overseas are among the most widely endorsed. Second, efforts aimed specifically at avoiding unintended pregnancy are popular, ranking much higher than those aimed at reducing population growth. Improving women's health and the status of women are also highly rated. Finally, reducing population growth rates is accepted and endorsed by most—but ranks below most other possible ways for involvement that we queried.

Three Types of Goals

Factor analysis shows that the public clusters this group of fifteen goals into three sets. Within each set, people who rank one goal as high (or low) priority tend to rank the others similarly. The three sets are

- health-oriented and other humanitarian goals,

- international relations–oriented goals, and

- helping countries slow their rate of population growth.

Each of these is discussed below.

Health-Oriented and Other Humanitarian Goals. The first group (or factor) includes the 10 goals that are primarily humanitarian and health-oriented. These include aiming to improve children's, women's, and men's health, child survival rates, and women's status and economic conditions; protect the environment; promote human rights; relieve the level of suffering; and reduce unintended pregnancy. People who give high scores to one of these goals are likely to support all the others in this set. In Appendix C, we show the order in which questions loaded onto this factor and we discuss the characteristics of the individuals who rank this factor (Factor 1A) highly. Individuals who are most enthusiastic about this type of goal are likely to be women, Democrats, African-Americans, and liberals.

Figure 3.5 illustrates how differently various demographic groups score the highest-ranking goal of "improving children's health." African-Americans and Hispanics are especially likely to give priority to the goal of improving children's health. Women and Democrats also rank this goal highly. Republicans are the least likely of all the groups shown in Figure 3.5 to rank this goal as a very high priority, although a quarter of Republicans rank this goal as a "10." This is higher than the overall averages for many of the other goals we asked about (see Figure 3.4).

We discuss support for specific health and humanitarian programs at the end of this chapter.

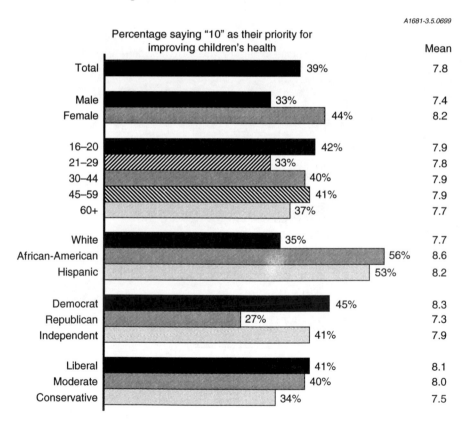

A1681-3.5.0699

Figure 3.5—What Percentage of Various Socioeconomic and Demographic Groups Assign High Priority to "Improving Children's Health"?

International Relations Goals. A second set of goals for assistance that grouped together in the factor analysis regards international relations: this includes promoting international trade, promoting democracy, supporting governments friendly to the United States, and preventing civil wars and conflicts. Figure 3.6 shows how different groups rate the goal of promoting trade. This is the goal that loaded first on the international relations factor in the factor analysis. Men, those age 45 or older, Hispanics, and African-Americans are most likely to rank the goal as a top priority, while women, younger people, and whites are the least likely to do so. Nonetheless, the mean figures of the 1–10-priority scores (shown on the right-hand side of Figure 3.6) exhibit remarkably little variation across subgroups.

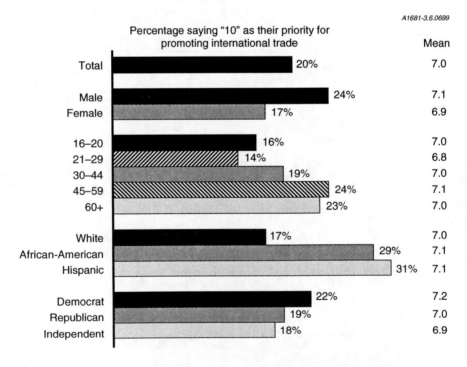

Figure 3.6—What Percentage of Various Socioeconomic and Demographic Groups Assign High Priority to "Promoting Trade"?

The Goal of Slowing Population Growth. Finally, the third set has only one goal in it: helping countries slow their rates of population growth. The factor analysis helps demonstrate that a desire for addressing population growth stands apart from the other goals. That is, the priority respondents assigned to this goal is not statistically related to the priorities they assigned to the other 14 goals for international economic aid. Figure 3.7 shows the scores given to the goal of slowing the rate of population growth by different groups.

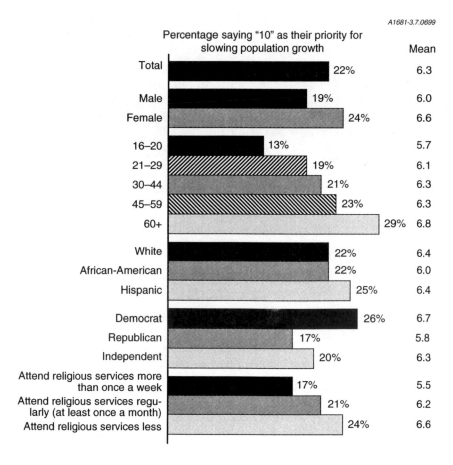

Figure 3.7—What Percentage of Various Socioeconomic and Demographic Groups Assign High Priority to Reducing Population Growth Rates?

The people most concerned with population growth are Democrats, less frequent religious service attendees, older Americans, and women. The relationship with age is particularly strong. Respondents aged 60 or older are more than twice as likely to assign a priority of "10" to slowing the rate of population growth compared to those aged 16–20 (29 percent vs. 13 percent).

VIEWS ON FUNDING PARTICULAR HEALTH AND HUMANITARIAN PROGRAMS

As seen above, many of the most highly rated goals for U.S. government assistance to other countries pertain to health and humanitarian matters. We asked respondents whether they specifically favored or opposed U.S. aid programs contributing to funding of particular types of health and humanitarian programs in developing countries that are part of the Programme of Action recommended by 178 countries at the International Conference on Population and Development that was held in Cairo in 1994.

Q32–39. Okay, now I'd like to ask your views about some kinds of possible U.S. aid programs to other countries. Please tell me if you favor or oppose the U.S. aid programs contributing to the funding of each of these in developing countries:

32. The U.S. sponsoring voluntary family planning programs in developing countries,

33. Programs to help women support themselves and their families financially,

34. Programs to give girls in developing countries the same opportunities for education as boys have,

35. Programs to improve women's health generally,

36. Efforts to reduce domestic violence against women,

37. Programs to encourage men to take an active role in practicing family planning,

38. Programs to improve the rate of survival of babies and young children,

39. Voluntary, safe abortions as part of reproductive health care in developing countries that request it?

The results are shown in Table 3.2. Of the eight programs listed, support is very high and strong for all but voluntary safe abortion as

Table 3.2

Support for Particular Health and Humanitarian Programs

	% Favor	% Oppose	% Favor Strongly	% Favor Some-what	% Oppose Some-what	% Oppose Strongly
Programs to improve the rate of survival of babies and young children	91	9	64	26	5	4
Programs to give girls in developing countries the same opportu-nities for education as boys have	90	9	72	18	5	5
Programs to encourage men to take an active role in family plan-ning	88	11	64	24	6	5
Programs to improve women's health gen-erally	88	12	52	36	6	5
Efforts to reduce domestic violence against women	85	14	63	22	7	7
Programs to help women support themselves and their families financially	84	15	55	29	8	7
The U.S. sponsoring voluntary family planning programs in developing countries	80	18	45	35	9	9
Voluntary, safe abor-tion as part of repro-ductive health care in developing countries that request it	50	47	24	26	13	33

part of reproductive health care in developing countries that request it. The factor analysis showed that people who supported one of the first seven programs listed tended to support the six others as well.

We discuss the results for voluntary family planning programs and abortion in more detail in Chapter Five.

Chapter Four

AMERICANS' KNOWLEDGE AND VIEWS OF WORLD DEMOGRAPHIC TRENDS

We have just looked at Americans' attitudes toward U.S. international economic assistance in general and their relative priorities for how it should be spent, with particular attention to the priority they assign to particular population-related goals vis-à-vis others. Before looking in more detail at the public's attitudes and opinions about specific population issues (in Chapter Five), we discuss in this chapter what respondents know about demographic trends and issues and how they view them.

KNOWLEDGE OF DEMOGRAPHIC FACTS AND TRENDS

First, we asked some general questions to test Americans' knowledge of global demographic facts and trends.

Q64. Could you give me an estimate of how many people there are in the world?

Americans' understanding of demographic facts and trends appears spotty. Although it acknowledges that world population is expanding, the public has little grasp of the dimensions of population size or growth (see Figure 4.1). Most do not know the size of the world population currently or the rate at which it is increasing, and they view the goal of slowing population as less urgent than other issues:

- Only 14 percent correctly identified the world's population as being in the 5 billion to 6 billion range. (The U.N. placed it at 5.9 billion in 1998, when the survey was fielded.)

A1681-4.1.0699

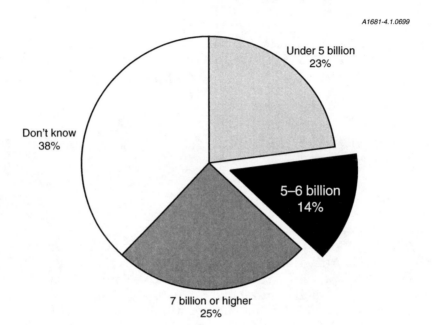

Figure 4.1—Give an Estimate of the World's Population

- The largest percentage of people (38 percent) say they do not know the world population size, a quarter underestimate it, and a quarter overestimate it.

- Fourteen percent of respondents think world population is more than 30 billion. In other words, the same number of people were off by a factor of five as knew the correct figure.

VIEWS OF POPULATION GROWTH RATES

Next, we asked about the rate of world population change.

Q65. To the best of your knowledge, would you say the world's population is growing, is at a stable level, or is shrinking?

The great majority (83 percent) of the public believes the world population is growing (see Figure 4.2). Only 13 percent think population has stabilized, and 3 percent say it is shrinking. Among voters, these percentages have not changed since 1994.

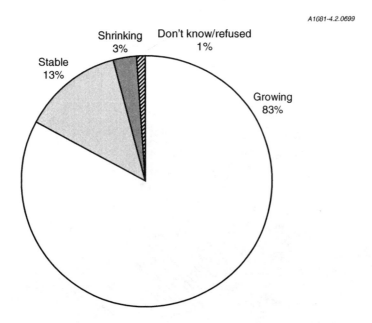

A1081-4.2.0699

**Figure 4.2—Is the World's Population Growing, Shrinking,
or Remaining Stable?**

We also asked those who said the world population was growing how long it would take for the world's population to double.

Q66. Please give me a rough estimate of how long you think it would take for the world population to double at current rates of growth?

Most Americans greatly overestimate how fast the world population is expanding. According to calculations of the UN Population Division, if existing rates were to continue, the 1998 population of 5.9 billion would double in 50 years (UN Secretariat, 1998). Even the most rapidly growing countries in the Middle East and Sub-Saharan Africa will take around 20 years to double in population (Population Reference Bureau, 1999). However, in the survey, three in ten (29 percent) respondents who thought the world population was growing said they thought it would double in 10 years or less, and two in ten (19 percent) in 11 to 20 years (see Figure 4.3). Another 11 percent place it at between 21 and 30 years. So a large majority believes the time it

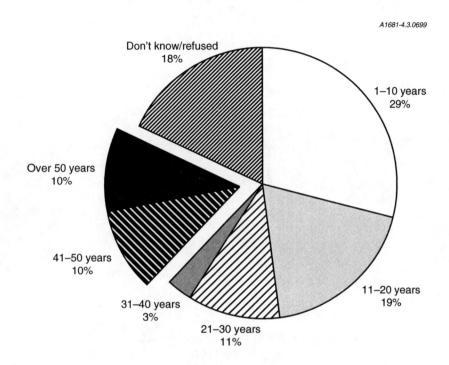

A1681-4.3.0699

Figure 4.3—How Long Will It Take for the World's Population
to Double at Current Rates of Growth?

will take for the population to double to be less than 30 years. Only
20 percent put doubling in the more realistic range of 41 to 50 years
(10 percent) or more than 50 years (10 percent). Eighteen percent
said they don't know.

*Q67. In your opinion, is the world overpopulated, underpopulated, or
would you say there is just about the right number of people in the
world? Is that very or somewhat?*

Nearly six in ten (59 percent) of the general public say the globe is
overpopulated. Twenty percent say it is very overpopulated, while 39
percent say it is somewhat overpopulated (see Figure 4.4). The
answers among voters on this question have not changed significantly since 1994, when it was also asked of the voting public.

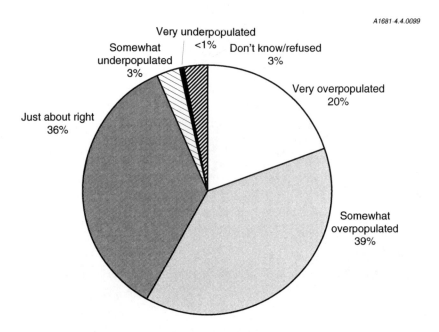

A1681-4.4.0099

Figure 4.4—Is the World Overpopulated, Underpopulated, or Just About Right?

HOW SERIOUS A PROBLEM IS POPULATION GROWTH?

Respondents viewed rapid population growth as a serious problem but less serious than most of the other international issues we asked about (see Figure 4.5).

Q10–14. Here are a few questions about the world more broadly. Using a scale where 1 means not at all a problem and 10 means it is a very serious problem, how big a problem do you think each of these international issues is?

10. Disease and hunger in other countries?

11. The spread of nuclear weapons?

12. Threats to the global environment?

13. Rapid population growth?

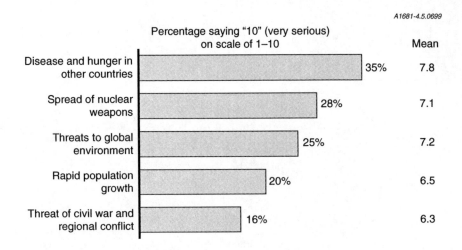

**Figure 4.5—To Which of These International Problems
Would You Assign High Priority?**

14. The threat of civil wars and international regional conflicts?

As in 1994, the American public is less concerned about rapid population growth than it is about disease and hunger, nuclear weapons,
and threats to the global environment (see Appendix A, Q10–14).
The percentage of voters who viewed population growth as an urgent
problem increased slightly since 1994, from 18 percent to 20 percent
but the difference is not statistically significant.

Older Americans and Hispanics are especially likely to view rapid
population growth as a serious problem (Figure 4.6). Those age 20 or
younger and those with a college education are least likely to hold
this view. These differentials are generally similar to those shown
earlier regarding the priority assigned to slowing population growth,
although the ethnic/racial differences shown in Figure 4.6 regarding
seriousness of population growth are much greater than those shown
earlier concerning priorities for reducing growth. This suggests that
opinions regarding the seriousness or importance of problems do
not necessarily translate into priorities for funding programs to
address these problems.

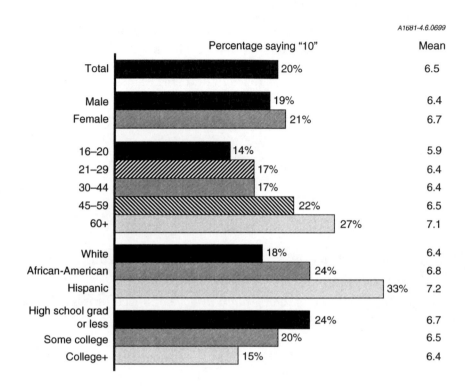

Figure 4.6—What Percentage of Various Socioeconomic and Demographic Groups Assign High Priority to "Rapid Population Growth"?

Our factor analysis finds that the seriousness of the four problems other than rapid population are viewed similarly by respondents and together form a factor (Factor 4—Global Problems), which is discussed in Appendix C. Rapid population growth does not group with these other problems but with a different set of questions about population issues.[1]

Q44. As I read each of the following statements, tell me whether you agree or disagree with it: population problems in the world have more to do with how people are concentrated in certain places, than with numbers of people. Is that strongly or somewhat?

---

[1]The three questions about the environment also grouped into a factor. See discussion of Factor 8 in Appendix C.

Most respondents (69 percent) said that population concentration is a more serious issue than population size per se (Figure 4.7). The proportion of voters agreeing with this statement is smaller in 1998 (68 percent) than it was in 1994 (75 percent). This is consistent with the slight (but not statistically significant) increase since 1994 we saw earlier in the percentage who viewed rapid population growth as a serious problem.

IMPACTS OF RAPID POPULATION GROWTH

The people of the United States tend to believe that population growth contributes to economic and other problems experienced around the world.

As I read each of the following statements, tell me whether you agree or disagree with it. Is that strongly or somewhat?

Q43. Too much population growth in developing countries is holding back their economic development.

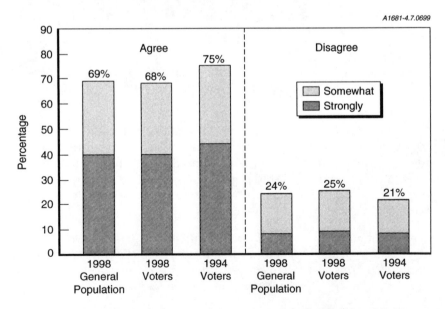

Figure 4.7—What Percentage of People Saw Population Problems as Arising from Concentration Rather Than from Numbers?

Q45. Rapid population growth in developing countries is frequently an underlying cause of civil war and regional conflicts.

Fully 71 percent of the general population in 1998 and 69 percent of voters agreed with the statement that "too much population growth in developing countries is holding back their economic development" (see Figure 4.8). This is markedly higher than the 55 percent figure for voters in 1994. The percentage of voters who strongly agreed with this statement in 1998 was twice the size of the percentage that strongly agreed in 1994.

Respondents are a little less sure about impacts on civil war and regional conflicts. Fifty-one percent agree with the statement that rapid population growth causes these problems, but 41 percent disagree (Figure 4.9). The responses for voters in 1998 do not differ significantly from those in 1994.

Opinions group together about rapid population growth holding back economic development and causing civil wars and regional

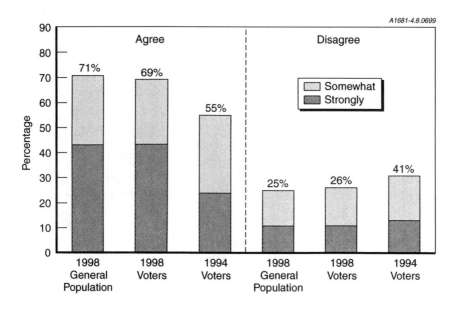

Figure 4.8—Does Excess Population Growth Hinder the Economies of Developing Countries? (1998 versus 1994)

A1681-4.9.0699

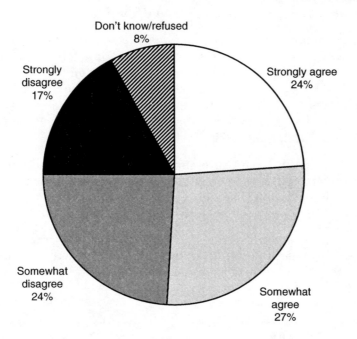

Figure 4.9—Is Rapid Population Growth Frequently an Underlying Cause of Civil Wars and Regional Conflicts in Developing Countries?

conflicts, the priority assigned to the goal of U.S. assistance for helping countries slow their rate of population growth, and the seriousness of population growth as a world problem. The differentials for "population growth holding back economic development," the question that loads first for this factor, are shown in Table 4.1. (Also, see discussion of Factor 5 in Appendix C.) Older Americans (age 60+) and Hispanics are especially likely to agree with the statement that population growth holds back economic development. African-Americans, Catholics, and those who attend religious services frequently are the most likely to disagree. Nonetheless, nearly two-thirds or more of every subgroup shown in the table agree with the statement.

Table 4.1

Opinions Regarding Whether Population Growth Holds Back Economic Development, by Socioeconomic and Demographic Characteristics

	% Agree	% Disagree	% Agree Strongly	% Agree Some- what	% Disagree Some- what	% Disagree Strongly
Total 1998	71	25	43	28	14	11
Gender						
Male	70	26	43	27	14	12
Female	71	23	42	29	14	9
Age						
16–20	74	23	40	34	13	9
21–29	71	26	38	33	18	7
30–44	69	29	39	30	16	13
45–59	68	27	42	26	14	13
60+	75	17	53	22	10	7
Race/Ethnicity						
White	71	24	43	28	14	9
African-American	65	31	33	31	15	15
Hispanic	75	25	51	24	14	12
Education						
High school grad or less	74	22	44	29	12	10
Some college	69	27	44	25	16	10
College+	68	27	39	29	16	11
Religious Affiliation						
High Protestant	77	19	46	31	13	6
Baptist	69	27	44	25	15	12
Other Protestant	72	24	45	27	13	11
Catholic	67	30	36	30	17	13
No religious affiliation	73	19	43	30	10	9
Religious Participation						
More than once a week	63	30	32	31	13	16
About once a week	71	25	42	29	14	11
Attend less	73	22	46	27	14	8

LOW FERTILITY IN DEVELOPED COUNTRIES

Sixty-one countries have already reached replacement fertility rates of 2.1 births per woman or lower, and many of the Northern Hemisphere nations will soon see an aged and even shrinking population, according to the latest biennial report of the U.N. Population Division (U.N. Secretariat, 1998). The argument has been put forth recently that low fertility in the developed nations is a greater problem than too many births in developing countries.[2]

Q48. As I read each of the following statements, tell me whether you agree or disagree with it. People in the developed, wealthier countries are having too few babies. Is that strongly or somewhat?

We found little agreement with this statement (see Table 4.2).

*   Two-thirds disagree with the assertion that "people in the developed, wealthier countries are having too few babies," while just a quarter agrees.

*   The assertion is disagreed with even more often by younger people, those with the most education and income, and those with no religious affiliation.

*   Hispanics show a tendency to agree more often than others do: 35 percent say there are too few births in developed countries; 25 percent agree strongly with this.

*   People age 60 and older (31 percent) and those with incomes of less than $25,000 per year (32 percent) are also among those most persuaded by this idea, yet even among those subgroups a large plurality disagrees with the statement that the wealthy nations have too few births.

*   However, in all categories, 49 percent or more disagreed with this sentiment.

Agreeing that too few babies are being born in the developed countries is associated (in factor analysis) with opposition to efforts enabling Americans to obtain family planning. That is, the small group of people who agree that lower fertility in developed countries is a problem are likely to disapprove of government funding to pro-

---

[2]See Wattenberg (1997). For a counterargument, see Population Action International (1998).

Table 4.2

Opinions Regarding Whether People in Wealthier Nations Are Having Too Few Children, by Socioeconomic and Demographic Characteristics

	% Agree	% Disagree	% Agree Strongly	% Agree Somewhat	% Disagree Somewhat	% Disagree Strongly	% Don't Know/ Refused
Total 1998	23	63	9	13	31	31	15
Gender							
Male	25	63	9	16	31	32	12
Female	21	62	10	11	31	30	17
Age							
16–20	23	68	9	15	39	29	9
21–29	19	73	6	13	35	37	8
30–44	21	65	9	12	32	33	14
45–59	20	62	8	11	32	30	18
60+	31	51	14	18	23	28	18
Race/Ethnicity							
White	21	65	6	14	33	32	14
African-American	24	61	13	11	28	33	15
Hispanic	35	52	25	9	25	27	14
Education							
High school grad or less	27	56	13	14	27	29	17
Some college	20	64	6	14	33	31	16
College+	20	71	8	12	36	34	10
Household Income							
<$25,000	32	49	13	19	21	28	19
$25,000–$49,999	21	65	9	12	34	31	14
$50,000–$74,999	19	72	7	12	38	34	10
>$75,000	22	67	10	12	33	34	11
Religious Affiliation							
High Protestant	23	65	9	15	33	32	11
Baptist	22	64	10	12	34	29	14
Other Protestant	22	59	11	11	27	31	19
Catholic	26	60	11	15	30	30	14
No religious affiliation	13	72	3	10	33	39	15

vide family planning services to the poor in the United States and to disagree that health insurers should cover family planning services as they do other doctor visits and coverage.

Chapter Five

SPECIFIC POPULATION ISSUES

This chapter discusses survey findings on four specific issues: views on the right to achieve desired family size, family planning, abortion, and congressional actions on population policy.

THE RIGHT TO CHOOSE AND ACHIEVE DESIRED FAMILY SIZE

We asked two questions in the survey about the right to decide and achieve desired family size.

Do you agree or disagree with the following statements. Is that strongly or somewhat?

Q46. People should feel free to have as many children as they can properly raise.

Q49. All couples and individuals should have the right to decide freely and responsibly the number, spacing, and timing of their children and to have the information and means to do so.

The first question had been asked in the 1994 survey. We asked it again in the 1998 survey so that we could see whether opinions changed during the interval. The second question asked about the principle underlying the ICPD Programme of Action.

Almost all Americans agree that people should be free to have as many children as they can properly raise, and half agree strongly (Figure 5.1). Among voters, 49 percent agree strongly, which is sig-

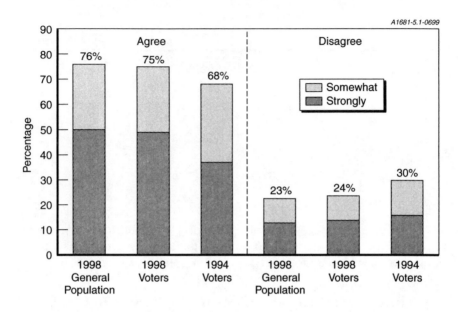

Figure 5.1—Should People Feel Free to Have as Many Children as They Can Properly Raise? (1994 Versus 1998)

nificantly greater than the 37 percent of voters who strongly agreed in 1994.

Nearly all respondents—92 percent—agree that couples should have the right to decide freely and responsibly the number, spacing, and timing of their children and to have the information and means to do so, and 69 percent agree very strongly (Figure 5.2). The percentage agreeing with this principle of reproductive freedom is at least 85 percent for every socioeconomic, demographic, political, and religious subgroup that we considered.

Responses to Questions 46 and 49 are highly correlated. The factor analysis reveals that endorsement of the principle of the right to decide the number and timing of births and the right to "have as many children as one can properly raise" are *not* associated with demographic goals or other health and assistance goals. People subscribe to these principles regardless of how they feel about supporting other aspects of the ICPD agenda or international economic

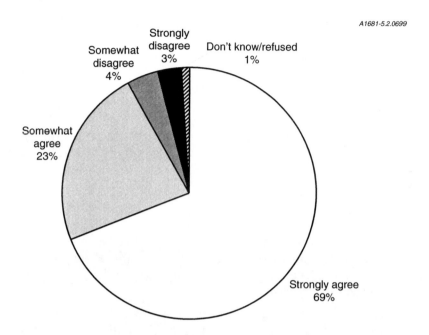

A1681-5.2.0699

Somewhat agree 23%

Somewhat disagree 4%

Strongly disagree 3%

Don't know/refused 1%

Strongly agree 69%

**Figure 5.2—Should People Have the Right to Determine Family
Size and Receive the Information and Means to Do So?**

assistance in general or how they feel about population growth. So
the high level of endorsement of the right to decide when and how
often to reproduce should not be interpreted as a high degree of concern for slowing population growth rates.

BELIEFS AND ATTITUDES ABOUT FAMILY PLANNING

Family planning programs have been a key policy mechanism used
to address population issues. Accordingly, we felt it was important to
understand the American public's definition of the core language
used to discuss these topics. We asked a series of questions about
the concepts of "birth control" and "family planning." These
explored respondents' perceptions of the meaning of the terms and
their views on the need for family planning, its availability both at
home and abroad, and support or opposition for U.S. funding of
international family planning efforts.

Meaning of the Terms "Birth Control" and "Family Planning"

"Family planning" and "birth control" are commonly used terms in all manner of media and communication.[1] We were curious about what the public perceives these terms to mean, and specifically if people consider abortion to be a part of the definition of either birth control or family planning. A random half of the respondents in the survey were asked to tell us in their own words what the term "birth control" means, and the other half what "family planning" means. The responses are shown in Table 5.1.

Q31a. Can you tell me what you think the term "birth control" means? What services does it include?

- To most respondents, "birth control" is, first, a technical term that describes the methods (e.g., "contraceptives") and results of using them ("preventing conception"); more than seven in ten offer definitions along these lines.

- Birth control has an educational meaning secondarily; 25 percent offer that the term means to give information or education.

- Birth control is also a behavioral concept to about a quarter of the public who say it means such characteristics as "being responsible" and "reducing unwanted pregnancy."

- Fifteen percent associate "birth control" with moderating population growth.

- Six percent mentioned abortion when asked about the meaning of birth control.

- Only 4 percent cannot or will not define "birth control."

Q31c. Can you tell me what you think the term "family planning" means? What services does it include?

---

[1]The term "birth control" was originated prior to World War I by Margaret Sanger in an effort to help legitimize the efforts of women to prevent pregnancy. In the mid-1930s, Sanger and others advanced the term "family planning" in an attempt to have reproductive health and pregnancy prevention included as part of the public health movement. The term "birth control" took on a new meaning with the introduction of modern contraceptives in the 1960s (Chesler, 1997).

"Family planning" seems to have a wider meaning than "birth control," one that includes decisions about whether to become pregnant, timing and spacing of children, and pregnancy prevention.

* To the largest number of people, it expresses a concept for behavior. Nearly half (48 percent) volunteered answers pertaining to such behavioral aspects as reducing unwanted pregnancy and having control and choice about pregnancy.

* Family planning also implies an educational quality (29 percent).

* It is a synonym for "birth control" for a large number (23 percent) as well.

* For a smaller number (16 percent), the term is associated with health and social well-being—chiefly prenatal and health care.

* For 15 percent the term "family planning" suggests methods of preventing pregnancy—such as contraception, abstinence, and sterilization—or abortion. Seven percent specifically mentioned abortion.

* Only 3 percent associate the term "family planning" with moderating population growth.

* A small portion of the public, 11 percent, defines "family planning" as unrelated to reproductive choice, methods, or related topics (e.g., family members planning a get-together), and 10 percent answered that they did not know or refused to answer. (Persons aged 16–20 were especially likely to answer "don't know" [23 percent] or to give an unrelated answer [28 percent]. That is, more than half of 16–20-year-old respondents did not or could not give an answer related to the term's common meaning. Only 7 percent of the youngest respondents replied "Don't know" for birth control, and none gave an unrelated answer.)

Do "Birth Control" and "Family Planning" Include "Abortion"?

As seen in Table 5.1, few people spontaneously mentioned that either "birth control" or "family planning" meant "abortion" (6 percent and 7 percent, respectively). After asking the open-ended ques-

Table 5.1

The Meaning of the Terms "Birth Control" and "Family Planning"
(Categorized Responses to Open-Ended Questions)

	A Birth Control (n = 750)	B Family Planning (n = 750)
"Family planning" (for Sample A)	3	N/A
"Birth control" (for Sample B)	N/A	23
Methods (net)	71	15
Contraceptives	42	6
Preventing conception	39	4
Abstinence, natural methods	8	1
Abortion	6	7
Sterilization	3	1
Education: Giving information, education	25	29
Behavioral concerns (net)	24	48
Reducing unwanted pregnancy, being responsible	11	25
Control over number of children conceived and born	2	20
Fewer children conceived and born	2	1
Having choice	9	7
Sexual freedom	1	0
Waiting until marriage to have children	1	3
Population growth (net)	15	3
Reduce population growth rate	14	3
Help foreign countries reduce population growth rate	1	1
Health care/Social benefits (net)	9	16
Prenatal care/health care	7	14
Reducing teen pregnancy	2	1
Miscellaneous related (net)	4	7
Adoption	0	1
Infanticide	1	0
Clinics	1	1
Other	3	6
Unrelated answers: a family event, planning for retirement, etc.	0	11
Don't know/Refused	4	10

Net = Total who gave a response in this category.

N/A = Not Applicable.

Because of multiple responses, percentages may add to more than 100%.

tions, we asked each subsample directly whether they thought the term included abortion.

Q31b.   When you hear the term "birth control," do you think it includes abortion?

Q31d.   When you hear the term "family planning," do you think it includes abortion?

When we asked specifically if each of these terms includes abortion, 33 percent said "birth control" includes it and 46 percent that "family planning" does (Table 5.2).

Both terms, especially "family planning," have multiple and different meanings for various people. In addition, both terms, especially family planning, imply abortion to many people. It is important not to assume the public interprets either of these uniformly. At the conclusion of the questioning about the meaning of family planning, respondents to the survey were given a definition of the term, which specified we were excluding abortion, as a reference for answering the subsequent questions concerning family planning. The statement read as follows:

For the purposes of this interview, I am going to use the term "family planning" and define it to mean having the information and services, including birth control or contraception, to determine if and when to

Table 5.2

Do "Birth Control" and "Family Planning" Include Abortion?

Q31b. When you hear the term "birth control," do you think it includes abortion?	
% Yes	33[a]
% No	66
% Don't know/Refused	1
Q31d. When you hear the term "family planning," do you think it includes abortion?	
% Yes	46[a]
% No	52
% Don't know/Refused	3

[a]Includes those who volunteered "abortion" as being included in the meaning of these terms in answering Q31a or Q31c.

*get pregnant and getting help with infertility problems. In this defini-
tion, family planning does not include abortion.*

Perceptions About Availability of Family Planning

We asked about perceptions of the availability of family planning
services.

*Q40. As far as you know, is family planning already available to most
people in all parts of the world today, or not?*

Only a small minority (18 percent) believes family planning is available in most parts of the world (see Figure 5.3). More than two-thirds
believe that it is not available to most people (and 14 percent either
don't know or refuse to answer). Lack of availability is perceived
most often by those individuals with the most education (see

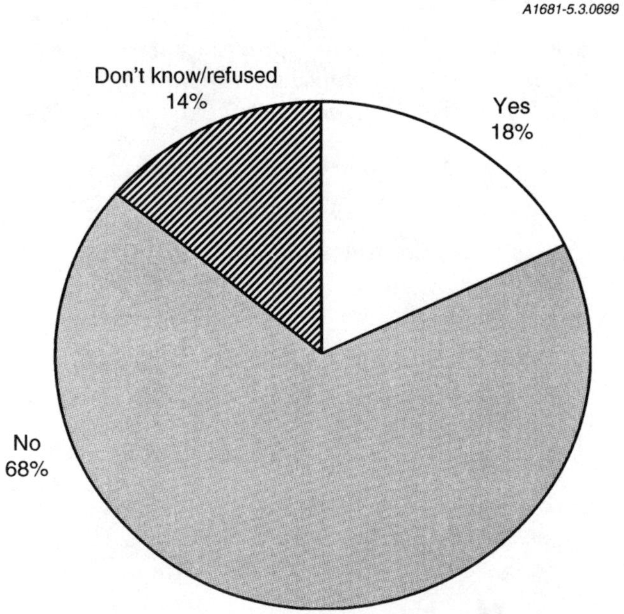

A1681-5.3.0699

Don't know/refused
14%

Yes
18%

No
68%

Figure 5.3—Is Family Planning Already Available to
Most People in the World?

Figure 5.4). Only 11 percent of those with a college education replied that it was available; 27 percent of those with a high school education or less said this is the case. More young people ages 16 to 20 (30 percent) believe family planning is available than any other group.

Next we asked about the availability of family planning services in the United States. The findings are very different from those pertaining to other parts of the world.

Q41. And as far as you know, is family planning already available to most people in the United States today, or not?

*   When thinking about the United States, a large portion of the public (84 percent) believes that family planning *is already* available to most people in this country (Figure 5.5).

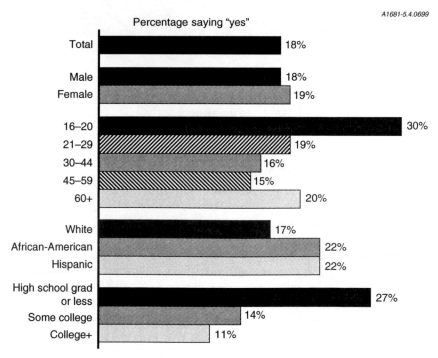

A1681-5.4.0699

Figure 5.4—Percentage of Various Socioeconomic and Demographic Groups Who Say Family Planning Is Already Available in All Parts of the World

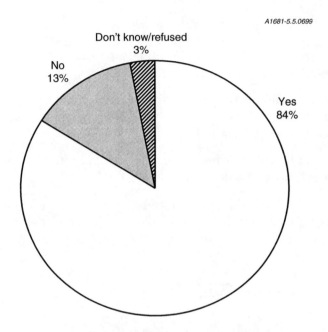

**Figure 5.5—Is Family Planning Already Available
to Most in the United States?**

• This percentage varies little by gender, age, or education, but African-Americans are less likely to think family planning is available than are whites and Hispanics (Figure 5.6).

Should Family Planning Be Provided by the Government and Covered by Health Insurance?

We asked two questions about family planning services in the United States. First, whether the government should fund family planning services for poor women as part of their health care, and, second, whether health insurance companies should cover family planning services as part of health care.

Q62. Thinking about here in the U.S., do you favor or oppose the government providing family planning services to poor women in this country who want them, as part of their health care? Is that strongly or somewhat?

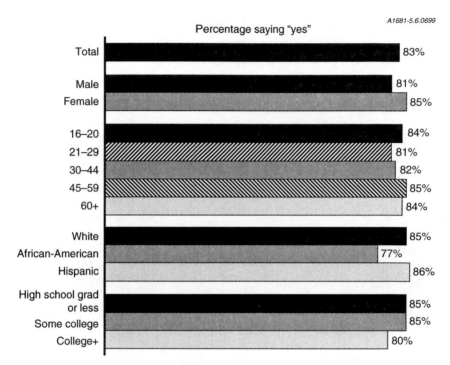

Figure 5.6—Percentage of Various Socioeconomic and Demographic
Groups Who Say That Family Planning Is Already
Available in the United States

On the domestic front, the vast majority (86 percent) of Americans
believe the government should provide voluntary family planning
services as part of poor women's health care, and most favor this
strongly. (Consistent with this belief, Congress has long provided
bipartisan support for domestic family planning.) At least three-
quarters of *all* of the subgroups shown in Table 5.3 favor such a program. Support is especially high among liberals, moderates, African-
Americans, and those ages 21 to 59.

Furthermore, the majority of the public believes that insurance companies should cover the cost of family planning.

*Q63. Do you agree or disagree that health insurers in the U.S. should
cover family planning services, as part of their regular health care
coverage? Is that strongly or somewhat?*

Table 5.3

Opinions Regarding Family Planning for Poor Women in the United States as Part of Government-Provided Health Care, by Socioeconomic and Demographic Characteristics

	% Favor	% Oppose	% Favor Strongly	% Favor Some-what	% Oppose Some-what	% Oppose Strongly
Total	86	13	56	30	6	6
Gender						
Male	84	15	51	33	7	8
Female	88	10	61	27	5	5
Age						
16–20	80	19	41	39	13	6
21–29	90	10	52	38	7	3
30–44	87	12	59	28	5	7
45–59	87	12	65	22	5	7
60+	82	13	53	29	6	7
Race/Ethnicity						
White	85	13	55	30	7	6
African-American	90	9	66	23	4	4
Hispanic	84	16	56	28	9	8
Political Viewpoint						
Liberal	90	10	66	23	6	3
Moderate	90	8	63	28	4	3
Conservative	80	19	44	36	8	11

Two-thirds agree strongly with this idea, and another 22 percent agree somewhat, for a total of 87 percent agreement. This sentiment is strongest among women and people under 60 years old, Democrats, liberals, and moderates and especially among minorities.

As with support for the government providing family planning services to poor American women who want them, the overwhelming majority of *all* subgroups shown in Table 5.4 support insurance coverage for family planning, and most do so strongly.

Support for U.S. Funding of International Family Planning Programs

As shown earlier when we discussed Q32–39, respondents were read a list of programs in developing countries and asked whether they

Table 5.4

Opinions Regarding Whether Health Insurers Should Cover Family Planning Services, by Socioeconomic and Demographic Characteristics

	% Agree	% Disagree	% Agree Strongly	% Agree Some-what	% Disagree Some-what	% Disagree Strongly
Total	87	12	64	22	6	6
Gender						
Male	83	15	57	26	8	8
Female	90	9	71	19	5	4
Age						
16–20	85	13	48	37	10	3
21–29	93	7	67	26	4	2
30–44	89	10	70	19	6	4
45–59	87	13	69	17	6	7
60+	78	18	55	23	6	11
Race/Ethnicity						
White	84	14	59	25	7	7
African-American	97	2	81	15	2	0
Hispanic	96	4	81	15	3	2
Party Affiliation						
Democrat	92	7	75	18	5	3
Republican	79	20	51	27	9	11
Independent	89	9	68	21	4	5
Political Viewpoint						
Liberal	93	6	75	18	4	2
Moderate	90	9	70	20	4	5
Conservative	80	19	53	27	10	9

favored or opposed U.S. aid to support them. In this section, we discuss the responses regarding voluntary family planning programs. For the purposes of this question, we clarified that we were defining family planning to exclude abortion.

Q32. Please tell me if you favor or oppose the U.S. aid programs contributing to the funding of each of these in developing countries: the U.S. sponsoring voluntary family planning programs in developing countries. Is that strongly or somewhat?

Eighty percent of respondents favor U.S. aid to support such programs, and 45 percent favor such support strongly. A substantial majority of *all* population subgroups favor support for funding for international family planning and most do so strongly. Hispanics are especially likely to voice strong support (Table 5.5).

In Figure 5.7 we present data on this question for respondents in the 1998 survey who voted in the most recent national election (i.e., 1996) and compare them to data in the 1994 survey (which interviewed only voters). As noted above, in the 1998 survey we clarified that for the purposes of this question we were defining family planning to exclude abortion. Such a clarification was *not* used in 1994.

In 1998, 78 percent of all voters favored U.S. aid to support family planning programs in developing countries, and 45 percent favored

Table 5.5

Support for U.S. Funding of Family Planning Programs in Developing Countries, by Socioeconomic and Demographic Characteristics

	% Favor	% Oppose	% Favor Strongly	% Favor Some- what	% Oppose Some- what	% Oppose Strongly
Total	80	18	45	35	9	9
Gender						
Male	76	22	41	35	11	11
Female	83	15	49	34	8	7
Age						
16–20	84	15	40	44	12	3
21–29	83	17	38	45	12	5
30–44	80	17	45	35	12	5
45–59	80	18	45	35	9	9
60+	75	21	45	28	6	15
Race/Ethnicity						
White	80	18	42	38	8	10
African-American	80	18	49	30	9	9
Hispanic	79	20	61	18	15	5
Party Affiliation						
Democrat	85	13	51	33	6	7
Republican	72	25	37	35	12	13
Independent	83	17	46	36	11	6
Political Viewpoint						
Liberal	84	15	51	33	8	6
Moderate	86	13	47	39	7	6
Conservative	72	25	39	34	12	13

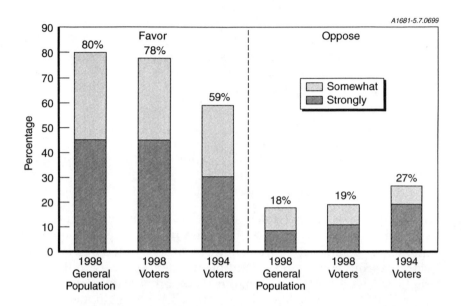

Figure 5.7—Should the United States Sponsor Voluntary Family Planning Programs in Developing Countries? (1994 Versus 1998)

such support strongly. These figures are considerably higher than those from the 1994 survey, when just under 60 percent of voters favored support for family planning programs in developing countries, and only 30 percent strongly favored such support. We see a 50 percent increase in the percentage who voice strong support. We do not know, however, whether the increase since 1994 represents a true increase in support for programs or whether our clarification in 1998 that we were defining family planning as excluding abortion contributed to the greater support in that year. To the extent that the latter is true, it suggests that the perception that these programs include abortion diminishes support for them.

Lack of support for U.S. funding of international family planning programs does not mean lack of support for domestic family planning (Table 5.6). Two-thirds of the 18 percent of the sample who oppose U.S. funding of family planning overseas favor the U.S. government providing such services to poor women in the United States who want them as part of their health care, although support is even greater for the 80 percent of the sample who favor support for family planning overseas (91 percent). Seventy percent of those who oppose support for family planning in developing countries feel that U.S. health insurers should cover family planning services as part of their regular health care (the percentage is even higher for those who favor support for international family planning [92 percent]).

ABORTION

Because discussions about family planning and population assistance are often influenced by the topic of abortion, we also asked a series of questions on that subject.

In stark contrast to the widely esteemed idea of family planning, abortion is an enduringly divisive issue in the American political landscape. It stands apart from views on other elements of women's issues, reproductive/family planning and health issues, and demographic concerns. As a nation, our views on abortion are divided, and people on each side of the abortion issue remain steadfast in their positions across many different lines of questioning.

Attitudes About Abortion in the United States

Attitudes about abortion in the United States have been remarkably stable over the last 25 years, but they come in many shades of gray

Table 5.6

Opinions About Domestic Family Planning, by Opinions About International Family Planning

	Among the 80% Who Favor U.S. Aid to Family Planning Programs in Developing Countries	Among the 18% Who Oppose U.S. Aid to Family Planning Programs in Developing Countries
Favor U.S. government funding of voluntary family planning services for poor American women (Q62)	91	67
Agree that health insurers in the United States should cover family planning services as part of regular health care (Q63)	92	70

between the far ends of the black-to-white spectrum. A long-used survey question asked opinions on abortion, offering three alternatives: should it be "legal under any circumstance, legal only under certain circumstances, or illegal in all circumstances." Figure 5.8 shows the stability in opinions elicited in response to this question in other surveys, with the exception of a rise in 1997 (as the late-term abortion issue arose) in the percentage saying they believe abortion should be legal only in some situations and corresponding drop in the percentage saying it should be legal under any circumstance.

The public is not divisible simply into pro- and antiabortion rights segments, because the great majority of Americans support or dis-

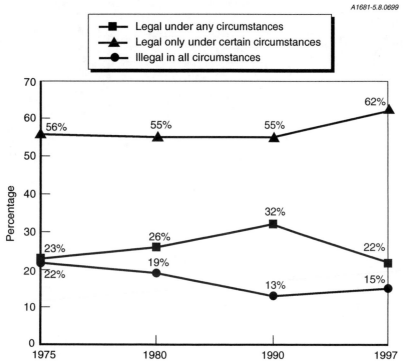

SOURCE: Surveys by the Gallup Organization for CNN/ *USA Today*, cited in *The Public Perspective: A Roper Center Review of Public Opinion and Polling*, December/January 1998.

Figure 5.8—Public Opinion Regarding Whether Abortion Should Be Legal or Illegal

approve of abortion depending on circumstances. The three-part
question is helpful because it has been used since 1975 and therefore
gives us a trend, but it is not a wholly satisfactory measure because of
the large number of people in the middle, circumstantial category.
In the current survey (Figure 5.9), we use a four-part question, which
has been used since 1995, to give a more exact measure of what the
public's views on this issue are (Arnedt, 1997).

*Q53. Thinking now about the abortion issue in the United States,
which of these comes closest to your view?*

Abortion should be generally available to those who want it;

*Abortion should be available but under stricter limits than it is
now;*

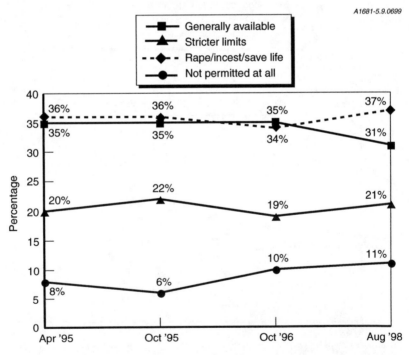

NOTE: Sources of the 1995 and 1996 data are general population surveys by CBS
News/ *New York Times,* cited in Arnedt (1997). The 1998 data are ours, using the
same question.

Figure 5.9—Public Opinion Regarding the Availability of Abortion

Abortion should be against the law except in cases of rape, incest, and to save the woman's life;

Abortion should not be permitted at all.

There is a shorter track record using the four-part question, but it too suggests that opinions on abortion have remained relatively stable (as shown in Figure 5.9) over the last few years.

The four-part question reveals that in 1998:

• about three in ten (31 percent) take the most liberal point of view, saying "abortion should be available to those who want it";

• about two in ten (21 percent) would let it be available but with stricter limits;

• 37 percent would restrict abortion to cases of rape, incest, or threat to a woman's life; and

• 11 percent would deny access entirely. (This is nearly double the 6 percent in 1995.)

Breaking the issue down, we can see that socioeconomic status, religious and political ideology all are associated with views on abortion issues (Table 5.7). Support is strongest among people who state no religious affiliation, those who do not attend church regularly, and liberals. By contrast, those who attend church more than once a week are especially unlikely to feel that abortion should be generally

Table 5.7

Attitudes on Conditions Under Which Abortion Should Be Available, by Socioeconomic and Demographic Characteristics

	% Generally Available	% Stricter Limits	% Only Rape, Incest, and Woman's Life	% Not At All
Total	31	21	37	11
Gender				
Male	32	20	37	11
Female	29	22	37	11

Table 5.7—continued

	% Generally Available	% Stricter Limits	% Only Rape, Incest, and Woman's Life	% Not At All
Age				
16–20	22	22	44	11
21–29	33	27	31	9
30–44	33	23	31	12
45–59	36	18	33	12
60+	24	16	50	9
Race/Ethnicity				
White	29	22	39	10
African-American	36	19	32	11
Hispanic	32	19	34	15
Education				
High school grad or less	24	17	46	12
Some college	30	20	37	12
College+	39	26	25	9
Household Income				
<$25,000	28	15	45	11
$25,000–$49,999	29	21	39	9
$50,000–$74,999	36	24	30	10
>$75,000	38	23	27	12
Party Affiliation				
Democrat	35	21	35	9
Republican	24	21	39	15
Independent	31	23	39	6
Political Viewpoint				
Liberal	44	23	27	6
Moderate	34	22	35	8
Conservative	19	19	45	16
Religious Affiliation				
High Protestant	37	22	36	5
Baptist	23	16	46	15
Other Protestant	25	19	39	16
Catholic	27	24	36	12
No religious affiliation	52	23	19	5
Religious Attendance				
More than once a week	11	12	44	30
About once a week	26	19	44	11
Attend less	42	25	27	5

available. The higher the education and income levels of a respondent, the more likely he or she is to support the liberal end of the abortion spectrum, and vice versa.

There is a similar rift in public opinion when the question is whether the government should fund abortion services for poor American women.

Q54. Do you favor or oppose the government providing funding for abortion services to poor women in this country who want them? Is that strongly or somewhat?

- Forty-seven percent favor and 51 percent oppose government funding for abortions for poor American women who want them (Table 5.8).

Table 5.8

Opinions Regarding Government-Funded Abortion for Poor Women in the United States, by Socioeconomic and Demographic Characteristics

	% Favor	% Oppose	% Favor Strongly	% Favor Some-what	% Oppose Some-what	% Oppose Strongly
Total	47	51	28	19	13	38
Gender						
Male	48	49	28	20	13	37
Female	46	53	28	18	14	39
Age						
16–20	50	48	24	25	20	29
21–29	47	51	26	21	17	34
30–44	48	50	26	22	11	39
45–59	49	50	32	16	10	40
60+	42	54	29	14	13	41
Race/Ethnicity						
White	45	53	26	19	14	39
African-American	63	36	34	28	10	26
Hispanic	46	53	33	12	11	42
Education						
High school grad or less	43	55	25	18	13	42
Some college	47	51	25	22	10	41
College+	53	45	35	18	15	30

Table 5.8—continued

	% Favor	% Oppose	% Favor Strongly	% Favor Some-what	% Oppose Some-what	% Oppose Strongly
Household Income						
<$25,000	46	51	27	20	12	39
$25,000–$49,999	45	53	26	19	14	39
$50,000–$74,999	48	50	31	18	12	39
>$75,000	56	42	38	18	9	32
Party Affiliation						
Democrat	56	43	33	23	13	29
Republican	36	62	23	13	14	48
Independent	49	50	28	21	14	36
Political Viewpoint						
Liberal	62	37	36	26	11	27
Moderate	52	46	32	20	17	29
Conservative	34	64	21	13	12	52
Religious Affiliation						
High Protestant	53	45	32	22	17	28
Baptist	39	59	22	17	13	47
Other Protestant	40	58	25	15	12	46
Catholic	45	54	25	20	12	41
No religious affiliation	66	32	42	23	11	21
Religious Attendance						
More than once a week	24	72	17	7	10	63
About once a week	44	55	24	20	15	40
Attend less	58	41	36	22	12	29

- Most people feel strongly one way or another about abortion funding for the poor—and opposition is especially intense: a robust 38 percent oppose government funding for abortion very much. This percentage rises to 63 percent among the 12 percent of the sample who attend church more than once a week.

- Support for government funding for abortion is especially high among African-Americans, liberals, and those who report no religious affiliation. (It is interesting that, although the question asks specifically about "poor" women, the lowest income group is *not* more supportive than others are.)

- Opposition to government funding for abortions for poor women is greatest and strongest among Republicans, conservatives, Baptists and "Other" Protestants, and those who attend church frequently.

Support for U.S. Funding of Abortions in Developing Countries

We also assessed opinions about support for U.S. funding of abortions in developing countries (see Table 5.9 and Figure 5.10):

Q39. How about voluntary, safe abortion as part of reproductive health care in developing countries that request it? Do you favor or oppose the U.S. aid programs contributing to the funding of this in developing countries? Is that strongly or somewhat?

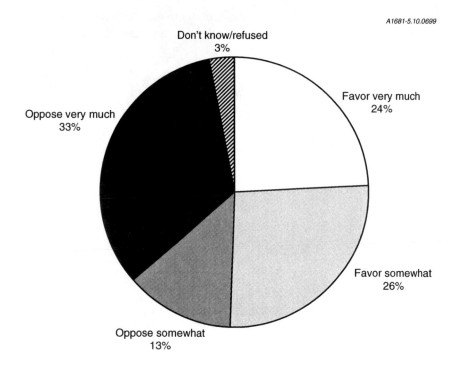

Figure 5.10—Do You Favor or Oppose U.S. Aid Programs Funding Voluntary Abortions in Developing Countries?

As we saw in Chapter Three, views are fairly equally divided on support for U.S. government funding of abortions overseas. Those who oppose such support tend to hold this view strongly. Liberals, High Protestants, those having no religious affiliation, and those who do not attend church regularly are more likely to support such funding, while those who attend church more than once a week are particularly likely to oppose such funding and to do so very strongly. Nearly two-thirds of those who attend church more than once a week "very much oppose" such programs. As with other questions we asked about abortion, Catholics express less opposition to abortion (49 percent oppose) than do Baptists (57 percent) and "Other Protestants" (54 percent).

Perceptions of Impacts of Abortion

The effects of abortion are the subjects of considerable debate in this country. We asked several questions to assess respondents' opinions about these effects.

Q55–58. Thinking about legal abortion in the U.S. and overseas, please tell me whether you agree or disagree with each of the following statements. Do you agree/disagree very much or somewhat?

55. Many women's lives are saved when abortion is legal, as compared to when abortion is illegal.

56. Too many women use abortion as a routine means of controlling births when it is legal.

Table 5.9

Support for U.S. Funding of Abortions in Developing Countries

	% Favor	% Oppose	% Favor Very Much	% Favor Some-what	% Oppose Some-what	% Oppose Very Much
Total	50	47	24	26	13	33
Gender						
Male	52	46	25	27	14	32
Female	49	47	23	26	13	34

Table 5.9—continued

	% Favor	% Oppose	% Favor Very Much	% Favor Some- what	% Oppose Some- what	% Oppose Very Much
Age						
16–20	52	45	22	29	16	29
21–29	56	41	19	36	17	24
30–44	52	46	22	30	12	33
45–59	51	47	31	20	11	36
60+	43	52	23	19	14	39
Education						
High school grad or less	46	50	21	25	14	36
Some college	52	45	25	27	11	35
College+	54	43	27	27	14	29
Party Affiliation						
Democrat	56	40	29	28	15	26
Republican	41	57	18	22	12	45
Independent	55	43	26	29	13	30
Political Viewpoint						
Liberal	62	35	32	30	12	23
Moderate	55	40	25	31	13	27
Conservative	39	59	18	21	15	45
Religious Affiliation						
High Protestant	60	37	28	33	13	24
Baptist	39	57	18	21	12	45
Other Protestant	44	54	19	25	15	39
Catholic	47	49	21	25	15	34
No religious affiliation	68	30	36	32	8	21
Religious Attendance						
More than once a week	24	72	13	11	9	63
About once a week	45	51	20	25	16	35
Attend less	63	34	31	32	12	22

57. Legal abortion encourages more sexual activity among teenagers and unmarried couples.

58. In most cases, women who have a legal abortion do so only as a last resort when their birth control fails.

We also asked a question similar to Q57 about the availability of contraception:

Q59. Do you agree or disagree that having legal contraception or birth control available encourages more sexual activity among teenagers and unmarried couples.

Most Americans consider legal abortion as lifesaving, but the public is ambivalent about whether it is used routinely as birth control by "too many women" and divided on whether the availability of abortion and contraception makes for more sexually active teenagers and unmarried people (see Table 5.10).

• Sixty-six percent say "many women's lives are saved when abortion is legal," and the majority of these strongly hold this view.

Table 5.10

Opinions Regarding Effects of Legal Abortion and Contraception

	% Agree	% Disagree	% Agree Very Much	% Agree Some-what	% Disagree Some-what	% Disagree Very Much
Many women's lives are saved when abortion is legal, as compared to when abortion is illegal.	66	29	41	26	14	14
Too many women use abortion as a routine means of controlling births when it is legal.	65	31	42	23	14	16
Legal abortion encourages more sexual activity among teenagers and unmarried couples.	53	45	35	18	17	28
In most cases, women who have a legal abortion do so only as a last resort when their birth control fails.	47	47	18	28	20	27
Having legal contraception or birth control available encourages more sexual activity among teenagers and unmarried couples.	49	48	29	20	19	29

- Sixty-five percent say "too many women" use legal abortion as routine birth control, and, again, this view is strongly held by many. However, Americans are split 47 percent to 47 percent when asked if they believe women who have abortions are doing so as a last resort, and respondents were more likely to strongly disagree with this statement than to strongly agree with it.

- Americans are also split on the questions of whether having legal abortion and contraception available leads to more teen and unmarried sex. Fifty-three percent say abortion has this result, and 45 percent disagree. Forty-nine percent say the availability of contraception has the same impact on behavior, and 48 percent disagree. Whichever position is taken, the views tend to be strongly held. It is interesting that respondents were more likely to think that legal abortion encouraged sexual activity among teenagers and unmarried couples than to think that legal contraception does this.

Correlates of Abortion

Factor analysis reveals that people have consistent opinions on abortion. Their answers to the various questions we asked on abortion (Q39, Q54–58), as well as the question (Q59) about whether legal contraception encourages more sexual activity among teenagers and unmarried couples, are closely related. For example, people who oppose support for U.S. funding for abortion in developing countries are much less likely to agree with the statement that many women's lives are saved when abortion is legal compared to when it is illegal, and they are more likely to feel that too many women use abortion as a routine means of controlling births and that legal abortion encourages sexual activity among teenagers and unmarried couples (and that the availability of contraception does this as well). People who oppose support for abortion in developing countries are also less likely than those who favor support for abortion abroad to feel that women only use abortion as a last resort when their birth control fails, and they are less likely to favor U.S. government support for abortion services for poor American women who want them.

Other than Q59 (regarding whether legal contraception encourages sexual activity among the unmarried and teenagers), respondents' opinions regarding abortion did not closely relate to their opinions regarding other matters asked about in the survey. For example, all other targets of international economic aid (Q32–38) clustered together into a single factor, but aid for voluntary safe abortion programs is not part of this factor.

The socioeconomic, demographic, political, and religious correlates of the abortion factor are similar to those discussed above for specific questions about abortion, with religious and political variables and education being particularly important. In addition, to the variables we have been discussing, Evangelical Christians and born-again Christians are especially likely to oppose abortion, while those who use the Internet and those who read newspapers, watch TV news, or listen to radio news frequently are more supportive of abortion. For a full discussion of the correlates of the abortion factor, see Appendix C (Factor 3).

RELATIONSHIP BETWEEN FAMILY PLANNING AND ABORTION

Since 1980, the public policy debate over whether the United States should fund international family planning programs has often been enmeshed in the issue of abortion. So we were particularly interested in how the American public perceives the relationship between family planning programs and abortion.

Q60. If family planning were made widely available in a country where it had not been, would you expect the number of abortions to fall, or to rise, or would having family planning widely available make no impact on abortion rates?

Half of Americans believe that providing family planning would reduce the number of abortions were it to be provided where it had not been previously available. Twenty-seven percent say it would have no impact, and 15 percent say making family planning available would cause abortion rates to rise (see Figure 5.11).

We also examined how opinions about U.S. funding for family planning programs in developing countries (Q32) relate to those about U.S. funding for voluntary safe abortion in countries that request it (Q39). The results of cross-tabulating responses to those two questions are shown in Figure 5.12.

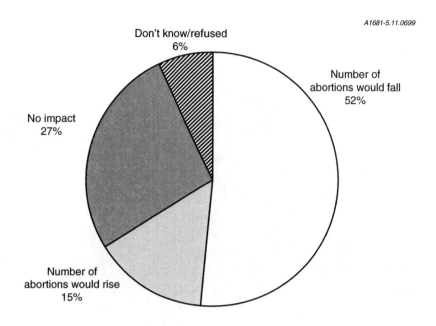

A1681-5.11.0699

Figure 5.11—Would the Number of Abortions Rise or Fall If Family Planning Were Made Available in a Nation Where It Had Not Been?

As seen earlier, eight out of ten respondents favor U.S. funding for voluntary family planning programs in developing countries, and half favor U.S. funding for voluntary safe abortion in countries that request it. The majority of those who favor funding for family planning programs in developing countries also favor support for voluntary safe abortion. Nearly all (more than 90 percent) who favor support for abortion also favor funding for family planning programs. Seventy percent of the 46 percent of the sample that oppose funding for abortion in developing countries nonetheless favor funding for family planning in those countries. Of our entire sample, 45 percent favor funding for both family planning programs and for abortion, 32 percent favor funding for family planning but oppose funding for abortion, while only 14 percent oppose funding for both.

We considered the three main subgroups shown in Figure 5.12— those who favor funding for both family planning programs and

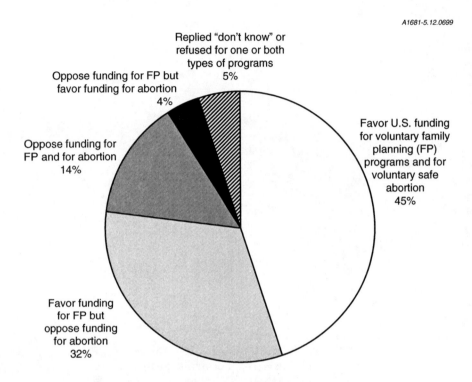

A1681-5.12.0699

Replied "don't know" or refused for one or both types of programs
5%

Oppose funding for FP but favor funding for abortion
4%

Oppose funding for FP and for abortion
14%

Favor U.S. funding for voluntary family planning (FP) programs and for voluntary safe abortion
45%

Favor funding for FP but oppose funding for abortion
32%

Figure 5.12—Do You Favor or Oppose U.S. Funding for Family Planning Programs and U.S. Funding for Abortions in Developing Countries?

abortion in developing countries, which we will call the "favor/favor" group; those who favor funding for family planning but oppose funding for abortion, which we will call the "favor/oppose" group; and those who oppose funding for both, which we will call the "oppose/oppose" group—and compared their characteristics and responses to those on other questions asked in the survey. We were interested in whether the group that favors funding for family planning but opposes funding for abortion (favor/oppose) looked more like others who favor funding for family planning or like others who oppose funding for abortion. Since some recent political debates about the funding of international family planning programs have tried to equate family planning with abortion, it is important to know more about the group that favors funding for family planning but not for abortion. Differences among these three groups for all key variables in the survey are shown in Table D.1 in Appendix D. We discuss some of the most interesting of those differences here. In comparing these groups in the discussion that follows, the reader should keep in mind that the groups differ considerably in size: the favor/favor group comprises nearly half of the sample (45 percent), the favor/oppose group comprises around one-third (32 percent), while the oppose/oppose group is only 14 percent.

In terms of their opinions about giving economic aid to other countries, the priority they assign to various goals for U.S. government assistance overseas, and support for specific programs, the favor/oppose group is very similar to the favor/favor group, while the oppose/oppose group gives lower priority to all of the goals and is less supportive of international economic aid in general and of all types of programs. For example, 65 percent of the favor/favor group and 66 percent of the favor/oppose group approve of economic assistance to other countries, while only 33 percent of the oppose/oppose group do. Thirty-one percent of both the favor/favor group and the favor/oppose group give the highest priority (10 on a 1–10 scale) to programs to promote human rights, while only 11 percent of the oppose/oppose group do; similar differences appeared for health programs for men, women, and children and most of the other programs we asked about. When asked about specific programs to help women and children in developing countries and to encourage men to take a more active role in family planning (Panel A of Table D.1), nearly all (92–98 percent) of both groups that favor funding for family planning favor funding for these programs, but typically less than half of the oppose/oppose group favor funding for such programs.

The favor/oppose group is somewhat less likely than the favor/favor group to feel that population growth is a serious problem, but the oppose/oppose group is especially unlikely to be concerned about population growth and its consequences. The favor/oppose group gives somewhat lower priority than the favor/favor group to helping countries slow their rate of population growth, is less likely to think that population growth in developing countries is holding back their economic development, is less likely to think that rapid population growth is frequently an underlying cause of civil war and regional conflict, and is less likely to think that the world is overpopulated.

However, the percentages of the favor/oppose group ranking these as the highest priority are all much higher than the corresponding figures for the oppose/oppose group. Both groups that favor support for family planning are about equally likely to think that the world population is growing (85 percent and 84 percent), while the oppose/oppose group is somewhat less likely to think this (75 percent).

The favor/oppose and oppose/oppose groups generally have similar attitudes about abortion. (See Panel E of Table D.1.) However, the group that favors funding for family planning but opposes funding for abortion resembles the group that favors both in expecting that making family planning available would reduce the number of abortions. Sixty percent of the favor/oppose group and 54 percent of the favor/favor group had this opinion, compared with only 29 percent of the oppose/oppose group. This suggests that an understanding of its potential for reducing abortion is associated with greater support for family planning.

Consistent with their support for family planning but not abortion, the favor/oppose group is the least likely of the three groups to use the terms interchangeably; they are the least likely to say that the terms "birth control" or "family planning" include abortion, both for the open-ended questions and when asked specifically.[2]

In terms of socioeconomic, demographic, and political characteristics, the two groups that oppose funding of abortion are similar in their high levels of frequent church attendance and the lower percentage identifying themselves as liberals. The favor/oppose group falls between the other two in terms of age—those who favor family planning tend to be younger, those who oppose abortion tend to be older—and political party affiliation and the percentages of moderates and conservatives. The favor/oppose group is the least educated of the three—48 percent have not gone beyond high school compared with 45 percent of oppose/oppose and 38 percent of

---

[2]It is interesting that the oppose/oppose group is especially unlikely to give an answer related to social and health benefits when asked the open-ended question. Only 2 percent of the oppose/oppose group mentioned health or social benefits when asked the meaning of birth control, compared with 8 percent for favor/oppose and 12 percent for favor/favor. The comparable statistics for the subsample that was asked the meaning of family planning are 10 percent, 16 percent, and 18 percent.

favor/favor—and its income is lower, but the differences by income are not great. The favor/oppose group contains the highest percentage of women of the three—56 percent—while the oppose/oppose group contains the lowest (44 percent).

VIEWS ON CONGRESSIONAL ACTIONS REGARDING POPULATION-RELATED FUNDING

In the last few years, Congress has made several efforts to limit U.S. support for family planning overseas and has continued to vote to withhold some of our dues to the United Nations. The public is fairly divided in its reaction to these efforts.

Q50–52. Please tell me if you approve or disapprove of each of the following positions Congress has taken:

- *Since 1995 Congress has voted to reduce the U.S. contribution to family planning in developing countries by 30 percent.*

- *Congress has voted to prevent the United States from funding family planning services in health care organizations overseas if those organizations also happen to perform abortions with other, non-U.S. funding.*

- Congress has withheld a portion of United Nations dues for the past 12 years.

Is that strongly or somewhat?

Table 5.11 presents answers to these questions.

The public is divided over reduction of U.S. contributions to international family planning programs. Fifty percent approve of the congressional vote to reduce the U.S. contribution to international family planning, and slightly fewer, 45 percent, disapprove of it. Neither approval nor disapproval are particularly strongly held opinions. The fact that a large number of Americans say they approve of Congress's cutting funding of family planning overseas belies the seeming support for providing such aid that we obtain in questions simply asking about support for types of aid without referring to levels of funding in the question.

Table 5.11

Opinions About Congressional Actions Regarding Funding of Family Planning and of the United Nations

	% Approve	% Dis- approve	% Approve Strongly	% Approve Some- what	% Dis- approve Some- what	% Dis- approve Strongly
Since 1995 Congress has voted to reduce the U.S. contribution to family planning in developing countries by 30%.	50	45	21	28	24	21
Congress has voted to prevent the United States from funding family planning services in health care organizations overseas if those organizations perform abortions with other, non-U.S. funding.	44	51	24	21	23	28
Congress has withheld a portion of U.N. dues for the past 12 years.	36	48	16	20	23	25

There is less approval of Congress's actions when we asked about preventing the United States from funding family planning in organizations that perform abortions—even if our contribution goes just for the family planning component of services. Here, 51 percent disapprove of Congress's denial of funding, and 44 percent approve; these views are more strongly held.

Withholding U.N. dues for the last 12 years is the least popular position of the three congressional actions we asked about: considerably more disapprove (48 percent) than approve (36 percent) of Congress's U.N. position. Fifteen percent of the sample either refused to answer the question about U.N. dues or replied that they didn't know. This is considerably greater than the corresponding figures of

5 percent and 4 percent for the other two congressional actions that we asked about.

Although public opinion shifts over these three questions, the factor analysis shows that there is a core group that answered the three questions similarly in support of Congress's actions to reduce or withhold funding. Regression analysis reveals that the core group that approves of these actions includes more men, conservatives and Republicans, while youth, people with higher levels of education, those who have traveled abroad for educational reasons, Catholics, and those with no religious affiliation tend to disapprove.

When we compare the favor/favor, favor/oppose, and oppose/ oppose groups (Panel H of Table D.1 in Appendix D), we see that those in the last group are especially likely to approve of the recent congressional actions, particularly reducing the U.S. contribution to family planning. Eighty-three percent of the oppose/oppose group approve of the reduction in the U.S. contribution to family planning. Although much lower than the 83 percent just mentioned, it is surprising that the percentage of those who favor funding for both family and abortion who do not disapprove of these funding cuts is substantial—40 percent.

Chapter Six

CONCLUSIONS AND IMPLICATIONS

Despite a perception that Americans have become more isolationist and absorbed with domestic problems since the end of the Cold War (see Kull [1996] and Murvachik [1995] for discussions of this perception), our survey found that most Americans feel that U.S. economic assistance to other countries is necessary and appropriate. Americans are somewhat concerned about population issues but less so than about a number of other issues. Furthermore, they lack knowledge about specific population facts and trends. Support for voluntary family planning activities is overwhelming. However, abortion remains a contentious issue that divides the American public.

This chapter summarizes and discusses the report's main findings and explores its implications for communicating population research.

INTERNATIONAL ECONOMIC ASSISTANCE AND RECENT CONGRESSIONAL ACTIONS

U.S. support for international economic assistance is at its highest level since 1974. Almost six in ten (59 percent) support U.S. economic assistance to other countries. Moreover, support for international economic assistance is 50 percent or greater for every socioeconomic, demographic, and political subgroup we considered.

Strong support appears for health-related and humanitarian programs, including those aimed at improving health, child survival, the environment, human rights, reducing suffering, and helping women in poor countries reduce unintended pregnancies.

Support is also high for programs that advance international relations goals, including promoting democracy, preventing war and conflict, promoting trade, and supporting friendly governments.

Support for helping countries slow population growth was not as high. The public believes that other phenomena, such as disease and hunger, the spread of nuclear weapons and threats to the world environment, are more serious problems. Our multivariate analysis also found that support for U.S. economic assistance to other countries is *not* associated with concern for population growth rates.

Although the American public does not put a high priority on funding population-related programs, it strongly subscribes to the principle of reproductive freedom for all. Support for this, the basic principle underlying the ICPD program, is universally high among all the subgroups interviewed.

The public has a mixed view of recent congressional actions regarding funding of international population-assistance programs. Fifty percent of the public approves of the 1996 congressional vote to reduce the U.S. contribution to international family planning, while 45 percent disapprove of it. The apparent contradiction between this response and the strongly held belief that the United States should support family planning programs overseas suggests either that respondents are unaware of historic or current funding levels or that Americans support such programs in principle but are less supportive when it comes to funding them.

There is less approval of Congress's actions in preventing the United States from funding family planning in organizations that perform abortions—even if the U.S. contribution goes only for the family planning component of services. Here, 51 percent disapprove of Congress's denial of funding and 44 percent approve.

Approval rates are even lower for Congress's withholding of a portion of the U.S. contribution to U.N. dues. Only 36 percent approved, while 48 percent disapproved.

SPECIFIC POPULATION ISSUES

Knowledge and Attitudes About Global Population Trends

The American public is only mildly interested in demographic issues and has a limited sense of the current dimensions of world population. Only 14 percent accurately report world population size in the five-to-six billion range. As many think it at least five times the current size as know the correct answer, and nearly 40 percent do not know. Furthermore, most show little grasp of the rates of growth. For example, nearly half say the world population will double in less than 20 years, when current demographic trends actually suggest that the doubling time will be around 50 years.

Notwithstanding this limited knowledge, a majority of respondents believe that the world is "overpopulated." A majority also believes that the concentration of population rather than growth per se is a problem and concurs with the assertion that population pressures contribute to holding back economic development in developing countries.

Family Planning

Most Americans see family planning programs as needed and beneficial. Eight in ten favor U.S. funding for voluntary family planning in developing countries. At least 70 percent of all demographic and socioeconomic subgroups we considered favor such funding. A majority believes that family planning is not available to most people in the world.

Very strong support (86 percent) exists for the government to provide family planning services to poor American women who want them; there is similarly strong support (87 percent) for requiring health insurers to cover family planning services for Americans. More than three-fourths of every demographic and political subgroup we considered favor such programs.

The majority (67–70 percent) of the relatively few who oppose funding for family planning programs in developing countries nonetheless favor government and insurers' support for family planning in the United States. Hence, the small opposition to funding family planning overseas does not seem to result from an opposition to family planning in general. Rather, it seems to spring from opposition to overseas economic assistance and perhaps also from a belief that rapid population growth is not a serious problem in developing countries.

Support for family planning in developing countries is related to a belief that it can reduce the number of abortions, to beliefs that the world is overpopulated and that rapid population growth is a serious problem, and to strong support for all types of humanitarian and other economic assistance to developing countries. Opposition to U.S. funding for family planning programs in developing countries seems to stem in part from a belief that the availability of contraception encourages sexual activity among teenagers and unmarried couples.

Abortion

Abortion remains an ever-present and divisive issue in the population policy arena. About half of the sample opposed abortion either completely or except in cases of rape, incest, and danger to the mother's life. U.S. attitudes about abortion have been fairly stable over the last 25 years.

Furthermore, two-thirds of respondents said that abortion is used by "too many" as a "routine means of controlling birth." About 47 percent said they believed that most women use abortion only as a last resort. Perhaps this apparent contradiction is explained by the possibility that opponents of abortion rights regard anyone's use of abortion as *"too many."*

Support for abortion appears to stem in part from a belief that legal abortion can save women's lives. Opposition to abortion appears to stem in part from belief that too many women use abortion as a routine means for controlling births and that the availability of legal abortion encourages sexual activity among teenagers and unmarried couples.

Attitudes about supporting abortion overseas are very similar to those regarding supporting abortion in the United States; those who oppose one are likely to oppose the other. Opposition is especially strong among Evangelicals, born-again Christians, those who attend religious services frequently, and among political conservatives.

Relationship Between Family Planning and Abortion

Half of the respondents agree with the statement that providing family planning would reduce the number of abortions were it to be provided where it had not been previously available. This belief is consistent with findings from demographic research, which has shown the potential for family planning to reduce abortion.

The majority (57 percent) of those who favor U.S. funding for family planning programs in developing countries also favor funding for voluntary safe abortion, and nearly all (90 percent) who favor funding for abortion also favor funding for family planning programs. In addition, more than two-thirds of those who oppose funding for abortion in developing countries support funding for family planning in those countries. Of our entire sample, 45 percent favored funding for both family planning and abortion, 32 percent favored funding for family planning but opposed funding for abortion, while only 14 percent opposed funding for both. Those who favor support for family planning but oppose support for abortion are similar to those who favor funding for both in their support for international engagement and in their belief that improved availability of family planning can reduce abortion, but they are more likely to align with the group that opposes funding for both on all other issues regarding abortion. This suggests that an understanding of the potential of family planning to reduce abortion is associated with support for family planning.

IMPLICATIONS FOR COMMUNICATING POPULATION RESEARCH

The survey findings have several implications for communicating the findings of population research to the public.

First, we found that very few people are aware of the size and rate of growth of the word's population. This finding suggests that many of the recent stories on "The Day of 6 Billion," in October of 1999 and afterward, might not have resonated with the public. Our survey suggests that this focus on aggregate numbers is less likely to interest the public than a focus on individual perspectives, such as helping women avoid unintended pregnancies. Stories about world population growing to "6 billion" seem to have little impact on the public, especially youth (who are even less concerned or informed about population size and rate of growth than the rest of the population). A focus on individual- and family-level quality-of-life issues, such as achieving desired family size, is consistent with the ICPD approach to framing population issues, although we are not able to assess whether the ICPD has had any causal effect on American attitudes.

Second, there is little linkage between the views people hold on the seriousness of population growth or its relationship to world problems and their views on reproductive rights, the environment, or other issues. Furthermore, the public cares less about population growth than it does about such issues as children's and women's health and the environment. Therefore, research communication could usefully emphasize the connections of population growth and high fertility with other issues Americans care about more. Research has shown, for example, strong links between women's fertility behavior and their own and their children's health.[1] Indeed, additional research that explores the intersections of these areas, cutting across traditional fields of analysis, would be valuable in advancing public understanding of how demographic concerns relate to other issues thought to be more pressing. The *Population Matters* project has several such projects under way: one, in publication, is examining the global security implications of demographic trends (Nichiporuk, forthcoming); the others, still in draft form, are examining the interrelations between demographic factors and environmental change and the relationship between population change and economic development.

---

[1] For an example of such research, see National Research Council, 1989.

Third, despite high levels of support for U.S. government funding for international family planning, half of the respondents did not oppose congressional cuts in funding for family planning programs in developing countries. Other research has shown that Americans tend to overestimate the fraction of the U.S. budget spent on international economic assistance. They might also do this for family planning programs. In fact, funding for family planning programs is about 4.5 percent of total international economic assistance, amounting to only pennies a day per American. The public would benefit from accurate information about the relatively low cost of population assistance programs and the need for—as well as governments' and individuals' continuing desire for—U.S. support for such programs.

Fourth, research shows that legal abortion can save women's lives, but only two-thirds of the overall population and only one-half of those who oppose U.S. support for abortion overseas recognize this.

Fifth, the public could be better informed about what research has shown regarding the potential of family planning services to reduce the prevalence of abortion. Research from a number of countries[2] shows that improved availability of contraception has cut the number of abortions.

Finally, the public lacks a clear grasp of what the term "family planning" means and whether it encompasses abortion. This is not surprising, because the demographic research community itself does not seem to agree on a single definition of family planning. This finding implies that communicators should not always assume that their audiences know the meanings of terms like "family planning" or "birth control" and should define them whenever possible.

---

[2]For example, Bangladesh (Ahmed, Rahman, and van Ginneken, 1998), South Korea (Noble and Potts, 1996), Hungary (Balogh and Lampe, 1994), Russia (Popov, 1996), and Kazakhstan, Uzbekistan, and Kyrgyzstan (Westoff et al., 1998).

QUESTIONNAIRE WITH RESPONSE TOTALS

The interviews were conducted from August 12 through September 3, 1998. The sample size was N = 1,500 in U.S. population with an oversample of 200 of ages 16–20. Unless otherwise noted, all results are expressed in percentages. Percentages may add to 99 percent or 101 percent because of rounding. An asterisk (*) indicates less than 1 percent. A dash (—) indicates zero. The margin of sampling error for the total sample is ±2.5 percentage points for percentages and ±0.15 for means of 1–10 scales. Percentages and means are based on weighted data.

QUESTIONNAIRE

Hello, my name is ____ and I am an interviewer with a research firm, Belden Opinion Research. We are conducting a public opinion survey, talking with both older teens and adults. We are not selling anything; this is a confidential interview as part of a national survey of the public's attitudes. May I please speak with the person 16 years old or older, living in this household, who had the most recent birthday? [When respondent is on line, repeat intro as necessary.]

1. Have you ever traveled out of the United States on business, for your education, or for pleasure? [multipunch]	Yes, net	57
	Yes, business	14
	Yes, education	8
	Yes, pleasure	47
	No	43
	Don't know	—
	Refuse	—

2. Have you ever lived outside the U.S. for more than two months?	Yes	18
	No	82
	Don't know	*
	Refuse	—

Based on those who answer yes in questions 1 or 2, N = 887

3. How often have you traveled outside the U.S.—once, two to four times, five to ten times, or more than ten?	Once	21
	Two–four times	39
	Five–ten times	21
	More than ten times	17
	Don't know	2
	Refuse	*

4. Have you ever spent time in a developing country, such as those in Latin America, Africa, Asia, or the Caribbean?	Yes	25
	No	74
	Don't know	*
	Refuse	*

5. On a different topic, do you have access to the Internet at home or at work, whether or not you use it?	Yes	53
	No	47
	Don't know	*
	Refuse	—

Based on those who use the Internet, N = 818

6. How often do you use the Internet— never, rarely, one or two days a week, three or four days a week, or five to seven days a week?	Never	13
	Rarely	19
	One–two days a week	19
	Three–four days a week	16
	Five–seven days a week	33
	Don't know	*
	Refuse	*

7. Would you say that you read a daily or Sunday newspaper never, rarely, once or twice a week, three or four times a week, or five to seven times a week?	Never	8
	Rarely	13
	One–two days a week	21
	Three–four days a week	13
	Five–seven days a week	45
	Don't know	*
	Refuse	*

8. Would you say that you watch or listen to a news program on television or radio never, rarely, once or twice a week, three or four times a week, or five to seven times a week?

Never	2
Rarely	5
One–two days a week	9
Three–four days a week	15
Five–seven days a week	70
Don't know	*
Refuse	*

9. Now I'd like your opinion on some issues. Do you think things in this country are going in the right direction or are things off on the wrong track?

Right direction	48
Wrong track	43
Don't know	8
Refuse	1

Here are a few questions about the world more broadly. Using a scale where 1 means not at all a problem and 10 means it is a very serious problem, how big a problem do you think each of these international issues is?

	% saying 10 (Mean)		
	All	Voters	
	1998	1998	1994
10. Disease and hunger in other countries	35	32	30
	(7.8)	(7.7)	(7.7)
11. The spread of nuclear weapons	28	27	24
	(7.1)	(7.2)	(7.1)
12. Threats to the global environment	25	21	18
	(7.2)	(7.0)	(6.9)
13. Rapid population growth	20	20	18
	(6.5)	(6.6)	(6.6)
14. The threat of civil wars and international regional conflicts	16	16	18
	(6.3)	(6.4)	(6.9)

15. Are you generally in favor or opposed to the U.S. giving economic assistance to other countries? Is that very much (in favor/opposed) or somewhat (in favor/opposed)?

	% saying 10		
	All	Voters	
	1998	1998	1994
Favor very much	18	17	8
Favor somewhat	41	42	38
Oppose somewhat	22	21	28
Oppose very much	15	16	15
Don't know	3	3	5
Refuse	*	*	4

Thinking about where you would like to see the United States use its resources, here are some possible goals for U.S. government assistance programs to other countries. Using a scale where 1 means lowest priority and 10 means top priority, please rate these goals for U.S. government assistance overseas: the first one is: [rotate]

	% saying 10 (Mean)		
	All	Voters	
	1998	1998	1994
16. Promoting human rights	27	25	21
	(7.0)	(6.9)	(6.5)
17. Improving economic conditions in developing countries	15	15	9
	(6.3)	(6.3)	(5.8)
18. Improving the status of women	19	19	21
	(6.5)	(6.4)	(6.3)
19. Promoting democracy	25	28	23
	(6.9)	(7.2)	(6.9)
20. Relieving human suffering brought about by civil war and natural disaster	30	27	25
	(7.4)	(7.3)	(7.1)
21. Supporting governments that are friendly to the United States	20		
	(6.8)		
22. Protecting the global environment	36	32	26
	(7.7)	(7.5)	(7.0)
23. Preventing civil wars and regional conflicts	22	19	14
	(6.4)	(6.3)	(5.7)
24. Helping countries slow their rate of population growth	22	23	17
	(6.3)	(6.4)	(5.9)
25. Improving women's health	25		
	(6.9)		
26. Improving men's health	17		
	(6.3)		

	% saying 10 (Mean)		
	All	Voters	
	1998	1998	1994
27. Improving children's health	39 (7.8)		
28. Increasing survival rates of babies and young children	33 (7.4)		
29. Promoting trade between the United States and other countries	20 (7.0)		
30. Helping women in poor countries avoid unintended pregnancies	30 (6.9)		

Based on Split A = 750 respondents

31a. Can you tell me what you think the term "birth control" means? What services does it include?

Top mentions
Preventing pregnancy
(net) 71
Education (net) 25
Behavioral choices
(net) 24
Don't know/refuse *

Based on Split A = 750 respondents

31b. When you hear the term "birth control" do you think it includes abortion? (Includes those who mention abortion in 31a.)

Think it includes 33
Don't think it includes 66
Don't know 1
Refuse *

Based on Split B = 750 respondents

31c. Can you tell me what you think the term "family planning" means? What services does it include?

Top mentions
Behavioral choices
(net) 48
Methods (net) 15
Education (net) 29
Don't know/refuse *

Based on Split B = 750 respondents

31d. When you hear the term "family planning" do you think it includes abortion? (Includes those who mention abortion in 31b.)

Think it includes 46
Don't think it includes 52
Don't know 2
Refuse 1

For purposes of this interview, I am going to use the term "family planning" and define it to mean: having the information and services, including birth control or contraception, to determine if and when to get pregnant, and getting help with infertility problems. In this definition, family planning does not include abortion.

Okay, now I'd like to ask your views about some kinds of possible U.S. aid programs to other countries. Please tell me if you favor or oppose the U.S. aid programs contributing to the funding of each of these in developing countries: the first one is (32) the U.S. sponsoring voluntary family planning programs in developing countries? (Is that strongly or somewhat?)

	All	Voters	
	1998	1998	1994
Strongly favor	45	45	30
Somewhat favor	35	33	29
Somewhat oppose	9	8	17
Strongly oppose	9	11	20
Don't know	2	2	3
Refuse	*	*	*

How about: _____? Do you favor or oppose the U.S. aid programs contributing to the funding of this in developing countries? (Read each, repeat root question as needed.) (Is that strongly or somewhat?) (Rotate but 39 always last.)

	Strongly Favor	Some-what Favor	Some-what Oppose	Strongly Oppose	Don't Know/ Refuse
33. Programs to help women support themselves and their families financially	55	29	8	7	1
34. Programs to give girls in developing countries the same opportunities for education as boys have	72	18	5	5	1
35. Programs to improve women's health generally	52	36	6	5	1
36. Efforts to reduce domestic violence against women	63	22	7	7	1
37. Programs to encourage men to take an active role in practicing family planning	64	24	6	5	1
38. Programs to improve the rate of survival of babies and young children	64	26	5	4	1

	Strongly Favor	Some-what Favor	Some-what Oppose	Strongly Oppose	Don't Know/ Refuse
39. Voluntary, safe abortion as part of reproductive health care in developing countries that request it	24	26	13	33	3

40. As far as you know, is family planning already available to most people in all parts of the world today, or not?	Yes available	18
	No not	68
	Don't know	14
	Refuse	—

41. And as far as you know, is family planning already available to most people in the *United States* today, or not?	Yes available	83
	No not	13
	Don't know	3
	Refuse	*

As I read each of the following statements, tell me whether you agree or disagree with it. (Is that strongly or somewhat?) The first one is (rotate).

	Strongly Agree	Some-what Agree	Some-what Disagree	Strongly Disagree	Don't Know/ Refuse
42. Money we spend helping people overseas eventually helps the U.S. economically.	22	39	16	18	4
43. Too much population growth in developing countries is holding back their economic develop-ment.	43	28	14	11	5
	43[a]	26	15	11	5
	24[b]	31	28	13	4
44. Population problems in the world have more to do with how people are concentrated in certain places than with numbers of people.	29	40	16	8	7
	28[a]	40	16	9	7
	31[b]	44	13	8	5

	Strongly Agree	Some- what Agree	Some- what Disagree	Strongly Disagree	Don't Know/ Refuse
45. Rapid population growth in developing countries is frequently an underlying cause of civil war and regional conflicts.	24 26[a] 25[b]	27 27 32	24 23 25	17 16 12	8 9 6
46. People should feel free to have as many children as they can properly raise.	50 49[a] 37[b]	26 26 31	10 10 14	13 14 16	1 1 2
47. American overuse of resources is a major global environmental problem that needs to be changed.	43 41[a] 37[b]	30 30 40	14 14 12	9 11 8	4 3 3
48. People in the developed, wealthier countries are having too few babies.	9	13	31	31	15

[a]1998 voters; [b]1994 voters

49. Do you agree or disagree with the following statement: All couples and individuals should have the right to decide freely and responsibly the number, spacing, and timing of their children and to have the information and means to do so. (Is that strongly or somewhat?)	Strongly agree	69
	Somewhat agree	23
	Somewhat disagree	4
	Strongly disagree	3
	Don't know	1
	Refuse	*

Please tell me if you approve or disapprove of each of the following positions Congress has taken: (Is that strongly or somewhat?)

	Strongly Approve	Some- what Approve	Some- what Dis- approve	Strongly Dis- approve	Don't Know/ Refuse
50. Since 1995 the U.S. Congress has voted to reduce the U.S. contribu- tion to family planning in developing countries by 30%.	21	28	24	21	5

	Strongly Approve	Some-what Approve	Some-what Dis-approve	Strongly Dis-approve	Don't Know/ Refuse
51. Congress has voted to prevent the U.S. from funding family planning services in health organizations overseas, if those organizations also happen to perform abortions with other, non-U.S. funding.	24	21	23	28	4
52. Congress has withheld a portion of United Nations dues for the past 12 years.	16	20	23	25	15

53. Thinking now about the abortion issue in the United States, which of these comes closest to your view: a.) Abortion should be generally available to those who want it; or b.) Abortion should be available but under stricter limits than it is now; or c.) Abortion should be against the law except in cases of rape, incest, and to save the woman's life; or d.) Abortion should not be permitted at all? (Reverse order for alternate questionnaires.)

Generally available	31
Stricter limits	21
Only rape, incest, or to save life of the woman	37
Not permitted at all	11
Don't know	*
Refuse	*

54. Do you favor or oppose the government providing funding for abortion services to poor women in this country who want them? (Is that strongly or somewhat?)

Strongly favor	28
Somewhat favor	19
Somewhat oppose	13
Strongly oppose	38
Don't know	1
Refuse	*

Thinking about legal abortion in the United States and overseas, please tell me whether you agree or disagree with each of the following statements: (Do you agree/disagree very much or somewhat?)

	Very Much Agree	Some-what Agree	Some-what Disagree	Very Much Disagree	Don't Know/ Refuse
55. Many women's lives are saved when abortion is legal, as compared to when abortion is illegal.	41	26	14	14	5
56. Too many women use abortion as a routine means of controlling births when it is legal.	42	23	14	16	4
57. Legal abortion encourages more sexual activity among teenagers and unmarried couples.	35	18	17	28	2
58. In most cases, women who have a legal abortion do so only as a last resort when their birth control fails.	18	28	20	27	6

59. Do you agree or disagree that having legal contraception or birth control available encourages more sexual activity among teenagers and unmarried couples? (Do you agree/disagree very much or some-what?)

Very much agree	29
Somewhat agree	20
Somewhat disagree	19
Very much disagree	29
Don't know	2
Refuse	*

60. If family planning were made widely available in a country where it had not been, would you expect the number of abortions to fall, or to rise, or would having family planning widely available make no impact on abortion rates?

Number fall	52
Number rise	15
No impact	27
Don't know	6
Refuse	1

61. Do you agree more with those who say the United States should encourage developing countries to lower their birth rates, or more with those who say it is inappropriate for us to do this because it may offend other peoples' cultures?

	All	Voters	
	1998	1998	1994
Should encourage	42	43	55
Inappropriate	52	52	41
Don't know	5	5	4
Refuse	1	*	*

62. Thinking about here in the U.S., do you favor or oppose the government providing family planning services to poor women in this country who want them, as part of their health care? (Is that strongly or somewhat?)

Strongly favor	56
Somewhat favor	30
Somewhat oppose	6
Strongly oppose	6
Don't know	1
Refuse	*

63. Do you agree or disagree that health insurers in the U.S. should cover family planning services, just like other doctor's visits and services, as part of their regular health care coverage? (Is that strongly or somewhat?)

Strongly agree	64
Somewhat agree	22
Somewhat disagree	6
Strongly disagree	6
Don't know	1
Refuse	*

64. Next I have a few questions about the world's population. Could you give me an estimate of how many people there are in the world? (Do not read codes.)

Under one billion	6
1–2 billion	7
3–4 billion	9
5 billion	9
6 billion	6
7–8 billion	3
9–10 billion	2
11–30 billion	6
Over 30 billion	14
Don't know	38
Refuse	*

65. To the best of your knowledge, would you say the world's population is growing, is at a stable level, or is shrinking?

	All	Voters	
	1998	1998	1994
Growing	83	83	83
Stable	13	13	13
Shrinking	3	3	3
Don't know	1	2	2
Refuse	—	—	—

Based on those who say "growing" in question 65, N = 1,258

66. Please give me a rough estimate of how long you think it would take for the world population to double at current rates of growth? (Record number.) (Leave three spaces.)

1–10 years	29
11–20 years	19
21–30 years	11
31–40 years	3
41–50 years	10
Over 50 years	10
Don't know/refused	18

67. In your opinion, is the world overpopulated, underpopulated, or would you say there is just about the right number of people in the world? Is that very or somewhat over/underpopulated?

	All	Voters	
	1998	1998	1994
Very overpopulated	20	20	24
Somewhat overpopulated	39	41	36
Just about right	36	33	31
Somewhat underpopulated	3	3	3
Very underpopulated	*	*	2
Don't know	3	3	4
Refuse	*	*	*

Here are a few other questions about you.

D1. How often would you say you attend services in a church or synagogue or elsewhere—more than once a week, about once a week, at least once a month, a few times a year, less often than that, or never?

More than once a week	12
About once a week	32
At least once a month	15
A few times a year	22
Less often than that	9
Never	10
Don't know	*
Refuse	1

D2. What is your religious preference? (Do not read choices; if Protestant, ask: And what denomination would that be?)

High Protestant (net)	21
Baptist	21
Other Protestant (net)	20
Catholic	24
None (net)	9
Jewish	1
Muslim	1
Don't know	1
Refuse	2

Based on those who are Christians from question D2, N = 1,089

D3. Do you consider yourself to be a fundamentalist Christian?

Yes	37
No	46
Don't know	17
Refuse	*

Based on those who are Christians from question D2, N = 1,089

D4. Do you consider yourself to be a "born-again" Christian?

Yes	38
No	58
Don't know	3
Refuse	*

Based on those who are Christians from question D2, N = 1,089

D5. Do you consider yourself to be an Evangelical Christian?

Yes	16
No	71
Don't know	13
Refuse	*

D6. Thinking politically, do you consider yourself to be a Democrat, a Republican, an Independent, or something else? (Probe don't knows: Do you lean more to the Republican Party or more to the Democratic Party?)

Democrat	36
Republican	28
Independent	26
Something else	5
Don't know	4
Refuse	2

D7. When it comes to politics, do you consider yourself to be politically very liberal, somewhat liberal, a moderate, somewhat conservative or very conservative? (reverse order.)

Very liberal	6
Somewhat liberal	22
Moderate	28
Somewhat conservative	29
Very conservative	10
Don't know	4
Refuse	1

D8. Would you describe yourself as a feminist or not?	Yes	28
	No	67
	Don't know	5
	Refuse	1
D9. Which of the following best describes the place where you live: a large city, a suburb near a large city, a small city or town, or a rural area?	Large city	19
	Suburb	23
	Small city/town	38
	Rural area	20
	Don't know	*
	Refuse	*
D10. What was the last grade of school you completed?	Less than HS graduate	13
	HS graduate	30
	Some college	25
	College graduate	18
	Graduate work/ degree	13
	Don't know	*
	Refuse	*
D11. Are you married, divorced, separated, widowed, never been married, or living with someone else as a couple?	Married	52
	Divorced	10
	Separated	2
	Widowed	8
	Never been married	24
	Living with a partner	4
	Don't know	*
	Refuse	*
D12. Are you a parent?	Yes	69
	No	30
	Don't know	*
	Refuse	*

Based on those who have children from question D12, N = 1,002

D13. How many children do you have?	Mean	2.5
	Don't know	—
	Refuse	*
D14. Age	16–17	4
	18–20	6
	21–29	16
	30–44	31
	45–59	22
	60+	21
	Refuse	*

Based on those who are 18 or older in question D14, N = 1,389

D15.	Are you registered to vote at this address?	Yes	77
		No	23
		Don't know	*
		Refuse	*

Based on those who say they are registered to vote in question D15, N = 1,076

D16.	Did you happen to have a chance to vote in the election in November 1996 for President, when the candidates were Bill Clinton, Bob Dole, and Ross Perot?	Yes	85
		No	15
		Don't know	*
		Refuse	—

Based on those who say they voted in question D16, N = 910

D17.	For whom did you vote [rotate names]: Clinton, Dole, Perot, or someone else?	Clinton	50
		Dole	31
		Perot	8
		Someone else	2
		Don't know	2
		Refuse	6

D18.	Do you consider yourself Hispanic?	Yes	10
		No	90
		Don't know	*
		Refuse	*

D19.	And what is your race: Are you white, black or African-American, Asian or Pacific Islander, Native American, or something else?	White	83
		Black	12
		Asian	2
		Native American	1
		Something else	2
		Don't know	1
		Refuse	1
Combined race		White	75
		Black	11
		Hispanic	10
		Other	4

D20.	Stop me when I come to the category in which your household income fell before taxes in 1997: (read categories)	Less than $25,000	20
		$25,000–$49,000	35
		$50,000–$74,000	19
		$75,000–$99,000	8
		Over $100,000	7
		Don't know	4
		Refuse	6

That's all the questions. Thank you very much for your help.

Gender	Male	48
	Female	52
Region	Northeast	21
	Midwest	22
	West	20
	South	37

METHODOLOGY FOR THE SURVEY
AND THE ANALYSIS

SAMPLE

The universe for this study is all persons aged 16 and older living in the United States in telephone-equipped households. The sample was selected in two stages. In the first stage, the sampling frame was a list of randomly created phone numbers (a technique known as random digit dial, or RDD) for telephone exchanges in the United States created by SDR, Inc. (Sophisticated Data Research, Inc.). Telephone numbers were selected at random from this frame.

The second stage of sampling was selection at the household level. In residences where working telephones were reached, the survey respondents were selected using a random probability method; interviewers requested to speak with the person 16 years or older in the household who had the most recent birthday.

The survey contains a total of 1,500 completed interviews including an oversample of 200 young adults ages 16–20. The oversample was obtained via the same process as the larger sample. After we reached the target of 1,300 respondents over age 20, we asked the age of the respondent at the outset and only interviewed respondents age 16 to 20 years old.

The demographic characteristics of the sample, obtained via the selection methods just described, were matched to U.S. Census population estimates for the United States. The data collected in the survey have been weighted to adjust for differences between our sample and the U.S. population in age, gender, and race/ethnicity, to bring these demographic variables back to the proper proportion for the U.S. population and into their proper proportions within subgroups of age cohorts, racial/ethnic groups, and males and females. Table B.1 shows the unweighted and weighted percentages for the subgroups.

Table B.1

Composition of Population Groups of Respondents, Weighted to U.S. Census Totals

	Unweighted Number	Unweighted Percentage	Weighted Percentage
Total	1,500	100.0	100.0
White males 16–17	33	2.2	1.3
White males 18–20	33	2.2	1.9
White males 21–29	73	4.9	5.3
White males 30–44	173	11.5	11.2
White males 45–59	120	8.0	8.7
White males 60+	111	7.4	7.6
White females 16–17	32	2.1	1.2
White females 18–20	30	2.0	1.8
White females 21–29	70	4.7	5.2
White females 30–44	192	12.8	11.2
White females 45–59	160	10.7	8.9
White females 60+	149	9.9	10.2
Black males 16–17	7	0.5	0.3
Black males 18–20	6	0.4	0.4
Black males 21–29	19	1.3	1.0
Black males 30–44	15	1.0	1.8
Black males 45–59	7	0.5	1.0
Black males 60+	6	0.4	0.7
Black females 16–17	7	0.5	0.3
Black females 18–20	5	0.3	0.4
Black females 21–29	10	0.7	1.1
Black females 30–44	35	2.3	2.0
Black females 45–59	14	0.9	1.2
Black females 60+	11	0.7	1.1
Hispanic males 16–17	3	0.2	0.3
Hispanic males 18–20	9	0.6	0.4
Hispanic males 21–29	14	0.9	1.2
Hispanic males 30–44	16	1.1	1.8
Hispanic males 45–59	2	0.1	0.9

Table B.1—continued

	Unweighted Number	Unweighted Percentage	Weighted Percentage
Hispanic males 60+	2	0.1	0.5
Hispanic females 16–17	7	0.5	0.2
Hispanic females 18–20	13	0.9	0.4
Hispanic females 21–29	7	0.5	1.0
Hispanic females 30–44	12	0.8	1.7
Hispanic females 45–59	10	0.7	0.9
Hispanic females 60+	4	0.3	0.7
Other males 16–17	4	0.3	0.1
Other males 18–20	5	0.3	0.1
Other males 21–29	10	0.7	0.4
Other males 30–44	6	0.4	0.7
Other males 45–59	7	0.5	0.4
Other males 60+	3	0.2	0.3
Other females 16–17	4	0.3	0.1
Other females 18–20	2	0.1	0.1
Other females 21–29	7	0.5	0.4
Other females 30–44	8	0.5	0.7
Other females 45–59	4	0.3	0.5
Other females 60+	8	0.5	0.3
Refuse	13	0.9	0.1

SAMPLING ERROR

All sample surveys are subject to possible sampling error; that is, the results may differ from those that would be obtained if the entire population under study were interviewed. The size of sampling error depends on the number of interviews conducted. For variables presented as percentages, the margin of sampling error for a probability sample of 1,500 is plus or minus 2.5 percentage points at the 95% level of confidence. This means that in 95 out of 100 samples of this size the results obtained in the sample would fall in the range of plus or minus 2.5 percentage points of what would have been obtained if every person over 16 in the country had been interviewed.

The margin of sampling error for smaller subgroups within the sample will be larger. For example, the margin of sampling error for men (n = 688) is most conservatively estimated at plus or minus 3.8 percentage points, and for women (n = 812) is plus or minus 3.5 percentage points at the 95% level of confidence. Other nonsampling error, such as question-order effects or human error may also contribute to total survey error.

For the 1–10 scales, differences of 0.2 or more in mean scores for the total sample are statistically significant.

QUESTIONNAIRE AND INTERVIEWING

The questionnaire in this study was designed by Belden Russonello & Stewart (BR&S) and reviewed and approved by Sally Patterson of Wagner Associates, Ronald Hinckley of Research/Strategy/Management, and representatives of RAND, the David and Lucile Packard Foundation, and the Centre for Development and Population Activities (CEDPA). The questionnaire was then subjected to a pretest, resulting in further modifications in question wording and length. A copy of the survey questionnaire, along with response totals, is presented in Appendix A.

The 1998 survey repeated a number of questions asked in a 1994 survey for the Pew Global Stewardship Initiative. The Pew poll was conducted among individuals who said they voted in the prior (1992) presidential election. To track changes in attitudes over the four years we have provided the 1994 results and the directly comparable category of 1998 voters. The 1998 voters are those who said they voted in the prior presidential election, which, in this case, was 1996.

The fieldwork was conducted by telephone using a computer-assisted telephone interviewing system, from August 12 to September 3, 1998, by a team of professional, fully trained, and supervised telephone interviewers. A briefing session familiarized the interviewers with the sample specifications and the instrument for this study. The interviews averaged 21 minutes in length. BR&S monitored the interviewing and data collection at all stages to ensure quality and that the oversample of young adults was achieved.

FACTOR AND REGRESSION ANALYSES

Factor Analysis

The first step in the factor analysis was to take all of the "dependent" variables (questions 10–63) in the study and run a statistical procedure to group them according to their correlation or interrelatedness; the variables within each factor (a grouped set of variables) are more highly correlated with each other than with the variables in other factors. Variables that did not group with any other variables and were removed from the statistical model include questions 31, 40–41, and 44 (items that respondents did not answer in any fashion similar to how they answered other items). Question 31 is open-ended, asking respondents to define either "birth control" or "family planning" that did not necessarily have "right" and "wrong" answers. Questions 40–41 ask whether respondents believe that family planning is available globally and also in the United States. Question 44 asks respondents to agree or disagree with the statement, "Population problems have more to do with how people are concentrated in certain places, than with numbers of people." These "trees" did not fit into any classification scheme and their analysis is best performed in the item-by-item basic (cross-tab) analysis.

After removing the few nonfitting variables from the model, a final factor analysis was performed, resulting in 11 factors or underlying dimensions. These 11 concepts define the international economic assistance and population "forest" according to people's attitudes about these two areas and the domestic and international institutions associated with them. In this way, the 60 topical variables that we started with have been reduced to a more manageable set of 11 elements, which further describe how Americans view international economic assistance and other population issues.

The 11 factors are as follows:

1A. Humanitarian goals for international economic assistance.
1B. International relations goals for international economic assistance.
2. Targets of aid.
3. Abortion.
4. Global problems.
5. Population growth.
6. Congressional action.
7. Aid and payback.
8. Environment.

9. Low fertility in developed countries and
   family planning in the United States.
10. Right to decide.

The composition of each of these factors and the order in which the survey questions "loaded" on each are shown and discussed in Appendix C.[1]

Multiple Regression

We then conducted multiple regression analysis to identify which "explanatory" variables—be they demographic or behavioral—are most associated with each factor. Thirty demographic and behavioral variables (religion, politics, information input [news consumption and Internet], and travel experience) were regressed against each factor to determine the degree of their association, if any. This process also ranks the strength of association between predictor variables and controls for all predictor elements in each factor analysis. The results show the degree to which the factors divide or unify the population and where the divisions occur. They are discussed in Appendix C. Tables showing the actual regressions are available from the authors on request.

---

[1]Factor analysis compares each single variable against the combined weight of all of the variables. The single variable that has the strongest predictive capacity within the pattern is ranked first. That means it is loaded with the most weight.

DESCRIPTION OF THE 11 FACTORS AND
RESULTS OF REGRESSION ANALYSIS

FACTOR 1A: HUMANITARIAN GOALS FOR INTERNATIONAL ECONOMIC ASSISTANCE

Questions defining this factor:

Q25. Priority of U.S. government assistance to improving women's health.

Q27. Priority of U.S. government assistance to improving children's health.

Q28. Priority of U.S. government assistance to increasing survival rates of babies and young children.

Q26. Priority of U.S. government assistance to improving men's health.

Q18. Priority of U.S. government assistance to improving the status of women.

Q20. Priority of U.S. government assistance to relieving human suffering brought about by civil war and natural disaster.

Q30. Priority of U.S. government assistance to helping women in poor countries avoid unintended pregnancies.

Q16. Priority of U.S. government assistance to promoting human rights.

Q17. Priority of U.S. government assistance to improving economic conditions in developing countries.

Q22. Priority of U.S. government assistance in protecting the global environment.

"Humanitarian goals" is one of two factors where opinions on goals for international economic assistance group together. Formation of the two international economic assistance factors reveals what respondents believe are two thrusts of international economic aid; these groupings do not necessarily represent an endorsement of international economic assistance but rather are a statement of two areas in which the money should be spent if it is to be given.

This first factor is composed of questions that ask opinions on "humanitarian" questions, such as improving women's, men's, and children's health; increasing survival rates of babies; relieving human suffering; helping women avoid unintended pregnancy; promoting human rights; improving economic conditions; and protecting the global environment. Opinions about improving women's and children's health loaded first in this factor; however, there is still a strong degree of association with other nonhealth-relief efforts aimed at human rights issues, poverty, and the environment.

Regression analysis shows the individuals who are most enthusiastic about humanitarian goals for international economic assistance are women, Democrats, African-Americans, liberals, and those who have traveled to other countries for educational purposes.

FACTOR 1B: INTERNATIONAL RELATIONS GOALS FOR INTERNATIONAL ECONOMIC ASSISTANCE

Questions defining this factor:

Q29. Priority of U.S. government assistance in promoting trade between the United States and other countries.

Q19. Priority of U.S. government assistance in promoting democracy.

Q21. Priority of U.S. government assistance in supporting governments friendly to the United States.

Q23. Priority of U.S. government assistance in preventing civil wars and regional conflicts.

The "international relations" factor is the second of the U.S. assistance goal groupings. The international relations sentiments that hold together are promoting trade, promoting democracy, supporting U.S.-friendly governments, and preventing civil war and regional conflict. Promoting trade loaded first on this factor, and respondents' opinion on promoting trade showed strong correlation with opinions on more governmental questions.

Regression analysis shows that Democrats, older Americans, and those who are heavy television and radio news consumers are the most ardent supporters of the international relations type goals.

FACTOR 2: TARGETS OF AID

Questions defining this factor:

Q34. Favor/Oppose U.S. aid programs contributing to the funding of programs to give girls in developing countries the same opportunities for education as boys have.

Q37. Favor/Oppose U.S. aid programs contributing to the funding of programs to encourage men to take an active role in practicing family planning.

Q35. Favor/Oppose U.S. aid programs contributing to the funding of programs to improve women's health generally.

Q36. Favor/Oppose U.S. aid programs contributing to the funding of efforts to reduce domestic violence against women.

Q33. Favor/Oppose U.S. aid programs contributing to the funding of programs to help women support themselves and their families financially.

Q38. Favor/Oppose U.S. aid programs contributing to the funding of programs to improve the rate of survival of babies and young children.

Q32. Favor/Oppose U.S. sponsoring voluntary family planning programs in developing countries.

The formation of a factor out of seven of eight possible programs suggested by the ICPD Programme of Action indicates that people who support one program area are likely to support all. The only exception is abortion programs in countries where it is requested. Abortion does not correlate with the other elements of the ICPD Programme of Action but instead groups with other abortion items in Factor 3.

Regression analysis shows the targets of aid represented in Factor 2 were well-received in general but received even stronger support from younger Americans, liberals, females, Democrats, and frequent consumers of print media.

FACTOR 3: ABORTION

Questions defining this factor:

Q54. Favor/Oppose the government providing funding for abortion services to poor women in the United States.

Q39. Favor/Oppose U.S. aid programs contributing to the funding of voluntary, safe abortion as part of reproductive health care in developing countries that request it.

Q57. Agree/Disagree that legal abortion encourages more sexual activity among teenagers and unmarried couples.

Q55. Agree/Disagree that many women's lives are saved when abortion is legal, as compared to when abortion is illegal.

Q56. Agree/Disagree that too many women use abortion as a routine means of controlling births when it is legal.

Q58. Agree/Disagree that in most cases, women who have a legal abortion do so only as a last resort when their birth control fails.

Q59. Agree/Disagree that having legal contraception encourages more sexual activity among teenagers and unmarried couples.

The factor brings together opinions on abortion questions pertaining to reproductive health care, funding abortion for poor women, and the way that abortion is used and its relationship to sexual activity among teens. When people answer questions related to abortion, they answer consistently in a pro- or antiabortion manner; abortion stands out as a factor because people have consistent opinions on it as an issue.

While there are certainly differences of opinion on abortion among age and educational groups, regression analysis reveals that differences in opinions on abortion are more accurately predicted by religious, political, and informational variables. Attending religious services is the variable most strongly associated in regression analysis with the abortion factor. Most-frequent attendees of religious services are against abortion, while less-frequent attendees are more supportive of abortion rights. Attitudes on abortion are sharply divided by religious denomination as well: Jews and High Protestants (Methodists, Episcopalians, Presbyterians, and Lutherans) support abortion rights, while Evangelicals and born-again Christians and other Christian denominations (such as Pentecostal and Mormon) oppose them.

Politically, Democrats and liberals are more supportive of abortion than are Republicans and conservatives.

Also significant in the regression analysis of the abortion factor are information sources. Those who use the Internet and those who most frequently read newspapers are more supportive of abortion rights.

FACTOR 4: GLOBAL PROBLEMS

Questions defining this factor:

Q11. How serious a problem is the spread of nuclear weapons?

Q14. How serious a problem is the threat of civil wars and international regional conflicts?

Q12. How serious a problem are threats to the global environment?

Q10. How serious a problem is disease and hunger in other countries?

The "global problems" factor brings together opinions on how big a problem disease and hunger, nuclear war, threats to the global environment, and civil wars are: a somewhat "apocalyptic" grouping. Of note is that rapid population growth does not group with these variables but with a different set of questions about population issues.

Regression analysis shows women, African-Americans, frequent television and radio news media consumers, Christians who give Assembly of God, Mormon, United Church of Christ, and Pentecostal as their denomination, and youths agree most strongly that these are serious problems.

FACTOR 5: POPULATION GROWTH

Questions defining this factor:

Q43. Agree/Disagree that too much population growth in developing countries is holding back their economic development.

Q45. Agree/Disagree that rapid population growth in developing countries is frequently an underlying cause of civil war and regional conflicts.

Q24. Priority of U.S. government assistance in helping countries slow their rate of population growth.

Q13. How serious a problem is rapid population growth?

Opinions group together about rapid population growth holding back economic development, causing civil wars and regional conflicts, the goal of slowing population growth, and the seriousness of the population growth issue.

Regression analysis shows older respondents, those with lower levels of education, Hispanics, and those who attend religious services less often feel more strongly than average that increasing population pressures are related to conflicts and economic development. African-Americans, Catholics, and those with higher levels of education disagree that these crises are caused by rapid population growth.

FACTOR 6: CONGRESSIONAL ACTIONS

Questions defining this factor:

Q52. Approve/Disapprove of Congress withholding a portion of United Nations dues for the past 12 years.

Q50. Approve/Disapprove of Congress voting to reduce the U.S. contribution to family planning in developing countries by 30 percent since 1995.

Q51. Approve/Disapprove of Congress voting to prevent the United States from funding family planning services in health organizations overseas, if those organizations also happen to perform abortions with other, non-U.S. funding.

This factor is composed of opinions on the three questions that ask respondents' opinions about congressional actions in withholding funding from the United Nations or reducing or restricting funding for developing countries. Opinions about these actions of Congress cluster as their own group and do not associate with other variables.

Regression analysis shows that although public opinion shifts a lot over these three questions, there is a core group that answered the three questions similarly and affirmatively in support of Congress's actions. This core group is older, male, conservative, Republican, has lower levels of education, and has not traveled abroad for educational purposes. On the other side, opponents of the congressional positions tend to be young women, liberals, Democrats, those with more education, those who have traveled abroad for educational reasons, Catholics, and those with no religious affiliation.

FACTOR 7: AID AND PAYBACK

Questions defining this factor:

Q42. Agree/Disagree that money we spend helping people overseas eventually helps the United States economically.

Q15. Favor/Oppose U.S. giving economic assistance to other countries.

The "aid and payback" factor is based on attitudes toward two questions: the first one asks whether money spent overseas eventually helps us here in the United States, and the second one is favoring or opposing giving economic assistance. This factor brings together international economic assistance with self-interest—that helping others also helps at home. When people believe that the United States is getting something out of economic assistance, they are more likely to support that funding.

According to regression analysis, Americans with higher levels of education, youth, and those who have traveled abroad for pleasure are more likely to support aid that helps the United States, while Baptists, Lutherans, and those who attend church frequently are less supportive.

FACTOR 8: ENVIRONMENT

Questions defining this factor:

Q22. Priority of U.S. government assistance overseas for protecting the global environment.

Q12. How serious a problem are threats to the global environment?

Q47. Agree/Disagree that American overuse of resources is a major global environmental problem that needs to be changed.

Concern about U.S. overuse of global resources, the problem of disease and hunger, and a high priority for aid to protect the global environment group together to form this factor.

Opinions from three separate sections of the survey on environmental issues correlate very highly with each other.

Regression analysis reveals that support for environmental concerns was strongest among younger people, those with lower income levels, Democrats, women, liberals, and Americans with lower levels of education.

FACTOR 9: LOW FERTILITY IN DEVELOPED COUNTRIES AND FAMILY PLANNING IN THE UNITED STATES

Questions defining this factor:

Q48. Agree/Disagree that people in the developed, wealthier countries are having too few babies.

Q62. Favor/Oppose the government providing family planning services to poor women in this country who want them, as part of their health care.

Q63. Agree/Disagree that health insurers in the United States should cover family planning services, just like other doctor's visits and services, as part of their regular health care coverage.

Thinking that people in wealthier, developed nations are not having enough children correlates with opposition to providing family planning services for poor Americans and opposition to insurance companies covering family planning services in the United States.

Regression analysis shows that people over age 60 and those with incomes of less than $25,000 per year are the most persuaded by the idea that low fertility in developed countries is a problem; yet even among those subgroups a large plurality disagrees that the wealthy nations have too few births. This view is less endorsed by people under 30, those with the most education, income, and nonreligious individuals.

FACTOR 10: RIGHT TO DECIDE

Questions defining this factor:

Q49. Agree/Disagree that all couples and individuals should have the right to decide freely and responsibly the number, spacing, and timing of their children and to have the information and means to do so.

Q46. Agree/Disagree that people should feel free to have as many children as they can properly raise.

The opinions associated in the "right to decide" factor are having as many children as one can properly raise and the principle underlying the ICPD Programme of Action. The principle loads first in this factor, emphasizing the fundamental right to decide, while the association with Question 46 reinforces freedom.

These two questions enjoyed widespread support across constituencies, and as a result not many groups stand out as more strongly interested than others. However, African-Americans and women are somewhat stronger in their support than the general population.

DATA FROM RESPONDENTS WITH VARIOUS VIEWS ON ABORTION AND FAMILY PLANNING

Table D.1

Differences Among Respondents Who Favor U.S. Funding for Both Family Planning (FP) Programs and Abortion in Developing Countries, Favor Funding for Family Planning but Not Funding for Abortion, and Oppose Funding for Both

	Favor FP/Favor Abortion (n = 669)	Favor FP/Op-pose Abortion (n = 494)	Oppose FP/Op-pose Abortion (n = 199)
A. Foreign Aid			
Percentage favoring the U.S. giving economic assistance to other countries	65	66	33
Percentage who agree: Money we spend helping people overseas eventually helps the U.S. economically.	70	65	35
Percentage saying 10 (highest priority) for goals for U.S. government assistance over-seas for:			
Promoting human rights	31	31	11
Improving economic conditions in devel-oping countries	16	19	8
Improving the status of women	23	22	6
Promoting democracy	29	28	14

Table D.1—continued

	Favor FP/Favor Abortion (n = 669)	Favor FP/Oppose Abortion (n = 494)	Oppose FP/Oppose Abortion (n = 199)
Relieving human suffering brought about by civil war and natural disaster	32	35	18
Supporting governments that are friendly to the U.S.	22	22	13
Protecting the global environment	41	36	22
Preventing civil war and regional conflicts	23	26	8
Helping countries slow their rate of population growth	28	21	8
Improving women's health	28	27	12
Improving men's health	18	21	11
Improving children's health	40	46	20
Increasing the survival rates of babies and young children	35	40	17
Promoting trade between the U.S. and other countries	22	20	16
Helping women in poor countries avoid unintended pregnancies	36	34	12
Percentage who favor U.S. aid programs contributing to the funding in developing countries of:			
Programs to help women support themselves and their families economically	94	92	40
Programs to give girls in developing countries the same opportunities for education as boys have	97	96	57
Programs to improve women's health generally	97	95	45
Efforts to reduce domestic violence against women	94	94	42
Programs to encourage men to take an active role in family planning programs	98	98	42
Programs to improve the rate of survival of babies and young children	97	98	58
B. Views on World Population and Environment			
Percentage saying 10 (very serious) regarding seriousness of rapid population growth	24	17	16

Table D.1—continued

	Favor FP/Favor Abortion (n = 669)	Favor FP/Op-pose Abortion (n = 494)	Oppose FP/Op-pose Abortion (n = 199)
Percentage who agree:			
Too much population growth in developing countries is holding back their economic development	78	70	53
Population problems in the world have more to do with how people are concentrated in certain places than with numbers of people	69	73	63
Rapid population growth is frequently an underlying cause of civil war and regional conflicts	59	47	38
Americans' overuse of resources is a major global environmental problem that needs to be changed	76	76	63
Percentage who said world is overpopulated	63	57	45
Percentage who say it is inappropriate for us to enourage developing countries to lower their birthrates because it may offend other people's cultures	43	53	78
Percentage who say world population is			
Growing	85	84	75
Stable	12	12	15
Shrinking	2	3	6
Percentage who said family planning is not already available to most people in all parts of the world	74	65	62
C. Domestic Family Planning			
Percentage who said family planning is not already available to most people in the U.S.	16	11	10
Percentage who favor the U.S. government providing family planning services to poor American women who want them as part of their health care	93	88	59
Percentage who agree that health insurers in the U.S. should cover family planning services, just like other doctor's visits and services, as part of their regular health care	94	87	66

Table D.1—continued

	Favor FP/Favor Abortion (n = 669)	Favor FP/Op-pose Abortion (n = 494)	Oppose FP/Op-pose Abortion (n = 199)
D. Decisions About Childbearing			
Percentage who agree:			
People should feel free to have as many children as they can properly raise	72	82	76
All couples and individuals should have the right to freely and responsibly decide the number, spacing, and timing of their children and to have the information and means to do so	93	95	85
E. Opinions and Attitudes About Abortion			
Percentage who agree:			
Many women's lives are saved when abortion is legal, as compared to when abortion is illegal	82	54	42
Too many women use abortion as a routine means of controlling births	54	78	75
Legal abortion encourages more sexual activity among teenagers and unmarried couples	38	72	65
In most cases, women who have a legal abortion do so only as a last resort when their birth control fails	60	33	34
Having legal contraception or birth control available encourages more sexual activity among teenagers and unmarried couples	38	62	65
Percentage who favor government providing funding for abortion services for poor women in this country who want them	74	20	22
If family planning were made available in a country where it had not been, how would you expect the number of abortions to change?			
Fall	54	60	29
Rise	14	12	23
No Impact	24	23	39

Table D.1—continued

	Favor FP/Favor Abortion (n = 669)	Favor FP/Oppose Abortion (n = 494)	Oppose FP/Oppose Abortion (n = 199)
F. Meaning of "Birth Control" and "Family Planning"			
What does term "birth control" mean?			
Percentage who said abortion/abortion clinics (open ended)	6	2	9
Percentage who, when asked specifically, said "birth control" includes abortion	40	22	35
What does term "family planning" mean?			
Percentage who said abortion/abortion clinics (open ended)	7	5	13
Percentage who, when asked specifically, said "family planning" includes abortion	55	31	51
G. Socioeconomic, Demographic, Political, and Religious Characteristics			
Percentage who attend church at least once a week	33	56	54
Religious preference			
High Protestant	26	16	16
Baptist	16	27	22
Other Protestant	17	23	22
Catholic	23	26	27
Jewish	2	1	0
Muslim/Buddhist	1	1	1
No religious affiliation	11	5	7
Party Affiliation			
Percentage Democrat	41	34	23
Percentage Republican	23	30	46
Percentage Independent	28	25	22
Percentage who describe themselves as			
Liberal	36	21	22
Moderate	32	29	15
Conservative	29	45	60
Education			
Percentage with high school diploma or less	38	48	45
Percentage with some college	61	52	56
Age			
Percentage under 30	27	25	17
Percentage 30–59	56	51	58
Percentage 60+	18	23	25

Table D.1—continued

	Favor FP/Favor Abortion (n = 669)	Favor FP/Op-pose Abortion (n = 494)	Oppose FP/Op-pose Abortion (n = 199)
Percentage female	52	56	44
Percentage with annual household income:			
<$25,000	18	23	21
$25,000–$49,999	34	35	35
$50,000 or over	29	31	34
H. Opinions Regarding Recent Congressional Actions			
Percentage who approve of the following actions:			
Since 1995 the U.S. Congress has voted to reduce the U.S. contribution to family planning in developing countries	40	47	83
Congress has voted to prevent the U.S. from funding family planning services in health organizations overseas, if those organization also happen to per-form abortions with other non-U.S. funding	32	54	64
Congress has withheld a portion of U.N. dues for the past 12 years	33	35	53

BIBLIOGRAPHY

Ahmed, K., M. Rahman, and J. van Ginneken, "Induced Abortion in Matlab, Bangladesh," *International Family Planning Perspectives*, Vol. 24, No. 3, 1998, pp. 128–132.

Arnedt, Cheryl, CBS News Election & Survey Unit, "Measuring How the Public Feels about Abortion," paper presented at the American Association for Public Opinion Research, Norfolk, Va., 1997.

Ashford, Lori S., *Population Bulletin: New Perspectives on Population: Lessons from Cairo*, March 1995.

Balogh, A., and L. Lampe, "Hungary," in B. Rolston and A. Eggert, eds., *Abortion in the New Europe: A Comparative Handbook*, Westport, Conn.: Greenwood Press, 1994.

Belden, Nancy, "Survey of the American Public on the Third World, Interdependence, and Development Assistance," for the Overseas Development Council, Washington, D.C., July 1986.

Belden & Russonello, "A Report of Findings from a National Survey on Foreign Aid for the Rockefeller Foundation," Washington, D.C., February 1993.

Belden & Russonello, "Report of Findings from a National Survey on Population, Consumption, and the Environment," for the Pew Global Stewardship Initiative, Washington, D.C., March 1994.

Chesler, Ellen, "Margaret Sanger and the Birth Control Movement," in Paul A. Cimbala and Randall M. Miller, eds., *Against the Tide:*

Women Reformers in American Society, Praeger Press, Westport Conn., 1997.

Chicago Council on Foreign Relations, *American Public Opinion and U.S. Foreign Policy 1995*, 1995.

Doherty, Carroll J., "Foreign Policy: Is Congress Still Keeping Watch?" *Congressional Quarterly*, August 21, 1993, p. 2267.

Gould, Douglas, *Talking Population: A Practical Manual for Communications Professionals and Dedicated Colleagues in the Population Field*, Mamaroneck, N.Y: Douglas Gould & Co., Inc., 1997.

Keeter, Scott, and Carolyn Miller, "Consequences of Reducing Telephone Survey Nonresponse Bias or What Can You Do in Eight Weeks That You Can't Do in Five Days?" paper presented at the annual meeting of the American Association for Public Opinion Research, St. Louis, Mo., May 14–17, 1998.

Kull, Steven, "Americans and Foreign Aid," Program on International Policy Attitudes, March 1995.

Muravchik, Joshua, *The Imperative of American Leadership: A Challenge to Neo-Isolationism*, Washington, D.C.: The AEI Press, 1996.

National Research Council, *Contraception and Reproduction: Health Consequences for Women and Children in the Developing World*, Washington, D.C.: National Academy Press, 1989.

Nichiporuk, Brian, *The Security Dynamics of Demographic Factors*, Santa Monica, Calif.: RAND, MR-1088-WFHF/RF/A, forthcoming.

Noble, J., and M. Potts, "The Fertility Transition in Cuba and Federal Republic of Korea: The Impact of Organized Family Planning," *Journal of Biosocial Science*, Vol. 28, 1996, pp. 211–225.

Patterson, Sally, and David M. Adamson, *How Does Congress Approach Population and Family Planning Issues? Results of Qualitative Interviews with Legislative Directors*, Santa Monica, Calif.: RAND, MR-1048-WFHF/RF/UNFPA, 1999.

Popov, Andrei A., "Family Planning and Induced Abortion in Post-Soviet Russia of the Early 1990s: Unmet Needs in Information

Supply," in Julie DaVanzo, ed., *Russia's Demographic 'Crisis,'* Santa Monica, Calif.: RAND, CF-124, 1996.

Population Action International, "What Birth Dearth? Why World Population Is Still Growing," fact sheet, Washington, D.C., 1998.

Population Reference Bureau, "World Population Data Sheet," Washington, D.C., 1999.

United Nations Secretariat, Department of Economic and Social Affairs, *World Population Projections to 2150,* February 1998.

Wattenberg, Ben J., "The Population Explosion Is Over," *New York Times Magazine,* November 27, 1997.

Westoff, Charles, Almaz T. Sharmanov, and Jeremiah M. Sullivan, "The Replacement of Abortion by Contraception in Three Central Asian Republics," Washington, D.C.: Population Resource Center, 1998.

Yankelovich, Daniel, *Coming to Public Judgment: Making Democracy Work in a Complex World,* New York: Syracuse University Press, 1991.